GEOMETRY AND BEYOND WITH MATHOMAT™

The Geometer's Sketchpad™ and Concrete Materials

Michael O'Connor

OLM

OBJECTIVE LEARNING MATERIALS™

Innovation In Learning

ii

ISBN 978-0-9579405-5-0

First published 2013

Printed in Australia by
Snap Lilydale
Victoria 3140

Introduction

Geometry is one of the oldest areas of mathematics. It also runs as a continuous thread throughout the mathematics curriculum. From the very earliest encounters children have with shapes to the intricacies of formal proof, geometry is there.

This book is not intended as a comprehensive text on geometry, but rather as a collection of activities designed to guide students through the development of key concepts that link to the bigger picture. The activities lead students through an exploration of the properties with the intention that students discover many of the underlying concepts for themselves.

A number of the activities draw on Henri Picciotto's earlier work "Geometry Labs" published in 1999. I thank Henri for his inspiration and generosity in allowing these sections to be reproduced in this new work.

There are five sections, each dealing with a subset of the overall subject. Each section presents activities starting nominally at grade 5 and working up to year 10 level. Each lesson, with two exceptions, is connected to the Australian Curriculum by year level and Content Descriptor in a table on the first page of the activity. The two exceptions are extension activities in the Geometric Reasoning chapter that allow students to go deeper if time allows.

The selected concepts and the activities that explore them are connected together. If there is a gap in the knowledge or understanding of one concept then the ones that follow will be harder to grasp. It is, and always has been, the role of the teacher to consider what activities students need to explore in order to develop solid understanding of the material "in the curriculum". With this in mind teachers are encouraged to look at the activities and choose not just the ones "at level" but also the ones at earlier levels if they perceive that the students need them.

One of the emphases of the tasks is to develop communal understanding and consensus. Each lesson concludes with a discussion of the key points. Sometimes this is short, while at others it is quite detailed. In all cases, they are an integral part of the process of convincing the students that what they have discovered is true and does work.

Michael O'Connor

Acknowledgements

I would like to thank Henri Picciotto for his contributions to this work.
Pages 77-81, 113-130, and 135-168 of this book are based on Henri Picciotto's Geometry Labs,. which can be downloaded on Henri's website: www.MathEducationPage.org

Also thanks must be given to Kaye Stacey and Susie Groves for their comments and guidance in helping to develop a sense of how this book should be structured, and to Kristine Blacksell for trialling and reviewing early drafts of many of the activities.

Table of Contents

MEASUREMENT LESSON

1

Measuring Rectangles

Introduction

Measurement is about finding how large something is. For people, we can measure how tall we are, our height, and how heavy we are, our weight. With both of these measurements we can put together a picture of what a person might look like.

The same is true for two dimensional shapes like squares and rectangles. The two measurements that give us information about what a rectangle might look like are its PERIMETER and its AREA.

A.C. Level:	5
A.C. Ref No's:	Calculate the perimeter and area of rectangles using familiar metric units ACMMG109
A.C. Substrands	Measurement

Outcome

At the end of this activity

Students will know:

• How to calculate the perimeters and areas of various rectangles.

Students will be able to :

• Describe and draw rectangles when given values for perimeter and area.

Materials Required

Mathomat™, Grid paper

Activity 1

Defining Perimeter: The perimeter of a rectangle can be thought of as the length of string needed to completely wrap around the outside of the rectangle.

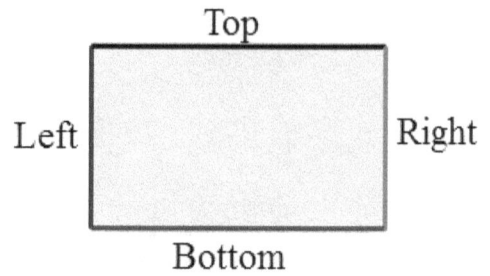

There are four sides to any rectangle: the Top, right, Bottom and Left. When these are lined up end to end against a ruler or tape measure we can read off the perimeter.

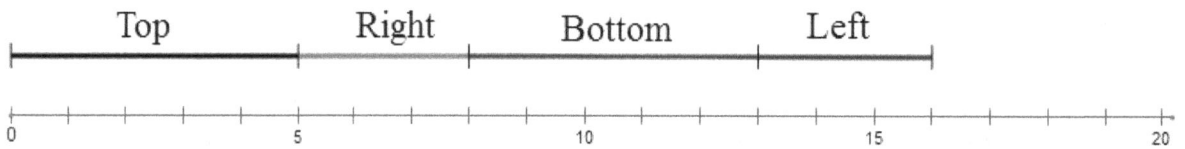

What is the perimeter of the rectangle above? _____

Another way to find the **perimeter** of a rectangle is to _____ the four side lengths together.

Defining Area: The area of a rectangle is the number of squares needed to fill up the inside of it. These squares are special in that they all have lengths and widths equal to one. They are called *unit squares*.

What is the area of the rectangle above?

Another way to find the **area** of a rectangle is to the length by the width.

For this rectangle, which measurement is larger, the **Perimeter** or the **Area**?

On the following two pages, there are several pictures of squares and rectangles. You need to calculate the perimeter and the area of each shape and write the answers in the spaces provided.

Squares

Square	Perimeter	Area
T L ⬚ R B	Top = ____ Right = ____ Base = ____ Left = ____ Perimeter = _____	Length: T = _____ Width: R = _____ Area =
T L ⬚ R B	Top = ____ Right = ____ Base = ____ Left = ____ Perimeter = _____	Length: T = _____ Width: R = _____ Area =
T L ⬚ R B	Top = ____ Right = ____ Base = ____ Left = ____ Perimeter = _____	Length: T = _____ Width: R = _____ Area =
T L ⬚ R B	Top = ____ Right = ____ Base = ____ Left = ____ Perimeter = _____	Length: T = _____ Width: R = _____ Area =
T L ⬚ R B	Top = ____ Right = ____ Base = ____ Left = ____ Perimeter = _____	Length: T = _____ Width: R = _____ Area =
T L ⬚ R B	Top = ____ Right = ____ Base = ____ Left = ____ Perimeter = _____	Length: T = _____ Width: R = _____ Area =

Rectangles

Rectangle	Perimeter	Area
T L R B	Top = ____ Right = ____ Base = ____ Left = ____ Perimeter = _____	Length: T = _____ Width: R = _____ Area =
T L R B	Top = ____ Right = ____ Base = ____ Left = ____ Perimeter = _____	Length: T = _____ Width: R = _____ Area =
T L R B	Top = ____ Right = ____ Base = ____ Left = ____ Perimeter = _____	Length: T = _____ Width: R = _____ Area =
T L R B	Top = ____ Right = ____ Base = ____ Left = ____ Perimeter = _____	Length: T = _____ Width: R = _____ Area =
T L R B	Top = ____ Right = ____ Base = ____ Left = ____ Perimeter = _____	Length: T = _____ Width: R = _____ Area =
T L R B	Top = ____ Right = ____ Base = ____ Left = ____ Perimeter = _____	Length: T = _____ Width: R = _____ Area =
T L R B	Top = ____ Right = ____ Base = ____ Left = ____ Perimeter = _____	Length: T = _____ Width: R = _____ Area =
T L R B	Top = ____ Right = ____ Base = ____ Left = ____ Perimeter = _____	Length: T = _____ Width: R = _____ Area =

Discussion

Questions and key points

For squares:

1) Under what conditions is the perimeter equal to the area?

2) Under what conditions is the perimeter greater than the area?

3) Under what conditions is the perimeter less than the area?

For rectangles:

1) Under what conditions is the perimeter equal to the area?

2) Under what conditions is the perimeter greater than the area?

3) Under what conditions is the perimeter less than the area?

Where to from here?

You have found that for both squares and rectangles there are three different ways in which perimeter and area can relate to each other. Is this also true for other shapes?

What shapes could you examine?

How could you go about finding the relationships that exist for these shapes? Will there always be three different groups?

MEASUREMENT LESSON

2

Measuring Many-Sided Figures

Introduction

A.C. Level:	6
A.C. Ref No's:	Solve problems involving the comparison of lengths and areas using appropriate units ACMMG137
A.C. Substrands	Measurement

Outcome

At the end of this activity:

Students will know:

- How to calculate the perimeters and areas of various closed polygons.

Students will be able to:

- Describe and compare polygons of different shapes when given values for perimeter and area

Materials Required

Mathomat™, Grid Paper

Activity 1

Two dimensional figures that:

- only have straight sides AND

- finish where they start (in other words are closed)

are called POLYGONS.

Poly means many and Gon means side, so all Polygons have many sides.

The simplest polygons are triangles, squares and rectangles.

All polygons have two properties that can be measured or calculated. These are Perimeter and Area.

In this activity we are going to work with polygons that are made up of collections of squares joined side to side with each other.

Calculate the Perimeter and Area of the square shown.

P =   A =

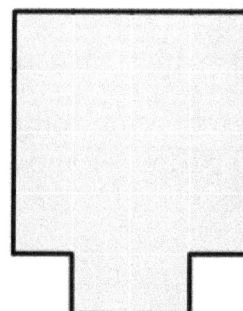

What do you notice about the two values?

Now consider the polygon with two extra 1 cm squares on the bottom of the original.

What are the perimeter and Area of this shape?

P =   A =

What do you notice about the values?

Can you draw at least three other polygons that have the same relationship between perimeter and area? Include the values of the perimeter and area for each figure.

What happens to the perimeter and area of any of the polygons you have looked at so far when you add a single 1 cm square to the diagram?

Try placing the square in different positions to see if the result is always the same or not.

Does this relationship perimeter and area exist for any other type of polygon that is not just a collection of squares? First let us consider triangles made by cutting rectangles in half along a diagonal. All side lengths are in cm.

What is the perimeter and area of each of these triangles?

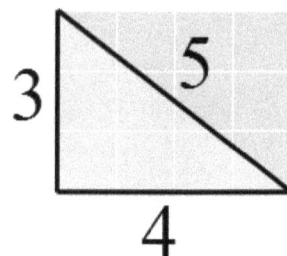

P =

A =

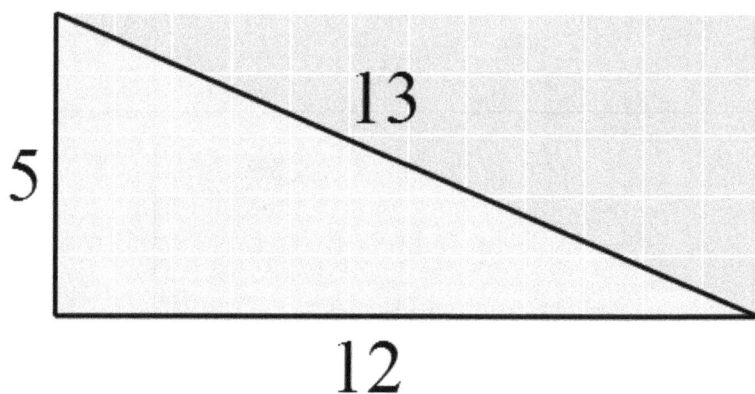

P =

A =

Which of these two triangles has the same relationship between perimeter and area as the polygons on the previous page?

Which do you think is more common, triangles with this property or ones without? Illustrate your answer with examples.

Now consider polygons of other types.

Take the two shown below as starting points. Verify that they too have the same relationship as the other polygons in this activity.

What are the lengths of the sloping sides? How do you know this is true?

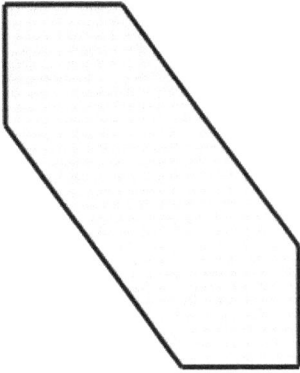

P =   A =   P =   A =

How many other polygons can you find that also have this property?

Try to find and draw at least two more of any sort.

Discussion

Questions and key points

Complete the following statement:

In this activity we have been working with polygons where the perimeter is _____ to the area.

Do all polygons have this property?

Can you describe the conditions for when a polygon is more likely to be of this sort and when it is not?

Where to from here?

All of the examples in this activity had perimeters that were even numbers. Is it possible to have a polygon with an odd perimeter that has the same relationship between perimeter and area as the ones in this activity? (Hint for teachers: side lengths do not have to be whole numbers.)

MEASUREMENT LESSON

3

Area Connections: Rectangles, Triangles and Parallelograms

Introduction

The unit of area is the square with side lengths of one unit each. Squares can be placed together to form rectangles and rectangles can be cut, folded and the parts shifted around to make triangles and parallelograms. In this lesson we will examine the connections between the areas of these three shapes.

A.C. Level:	7
A.C. Ref No's:	Establish the formulas for areas of rectangles, triangles and parallelograms and use these in problem solving. ACMMG 159
A.C. Substrands	Measurement

Outcome

At the end of this activity:

Students will know

*   How the areas of rectangles, triangles and parallelograms are related

Students will be able to

*   Calculate the area of each of these shapes.

Materials Required

Mathomat™

Activity 1

Cutting a rectangle in half along a diagonal gives a right angled triangle with half the area of the original triangle. Is this true for all triangles? Is it possible to take any triangle and find the rectangle that it was taken from, showing that every triangle is half of at least one rectangle?

Take the three triangles shown below.

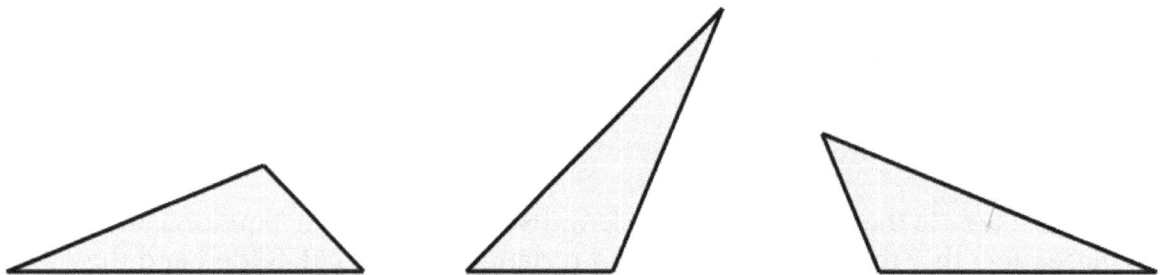

Work out the area of each triangle.

Draw a rectangle that you think the triangle was taken from.

Now work out the area of the rectangle.

Is the rectangle twice the area of the triangle inside it?

Discuss any differences with your group and compare your answers with theirs.

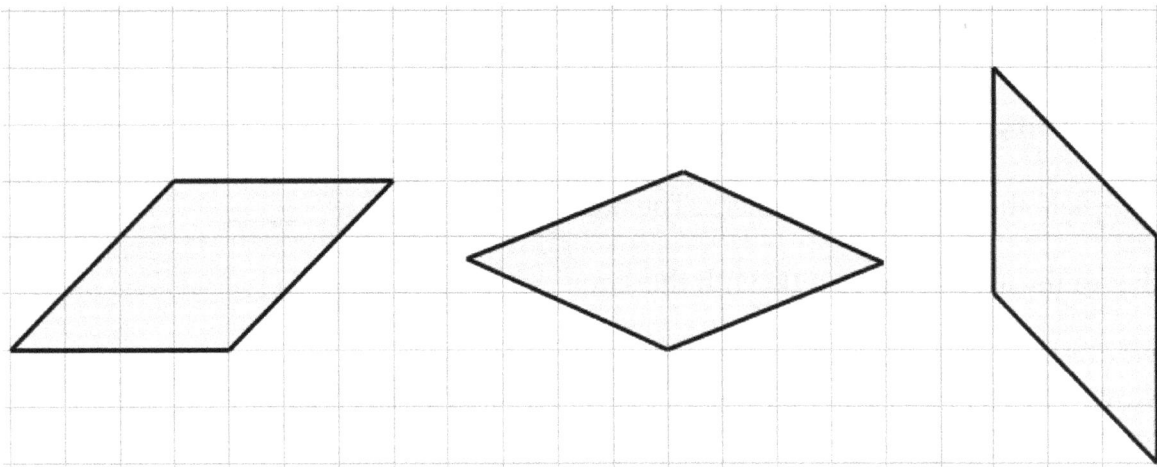

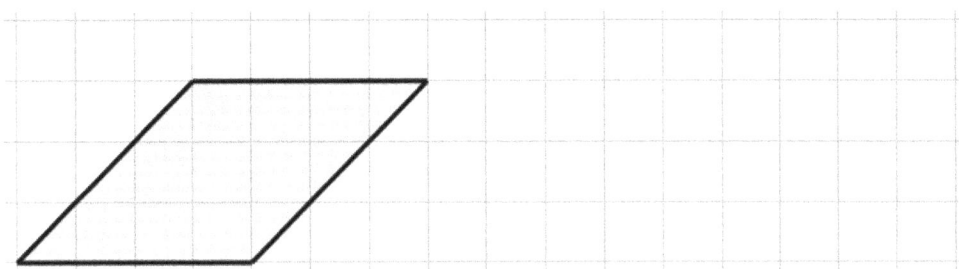

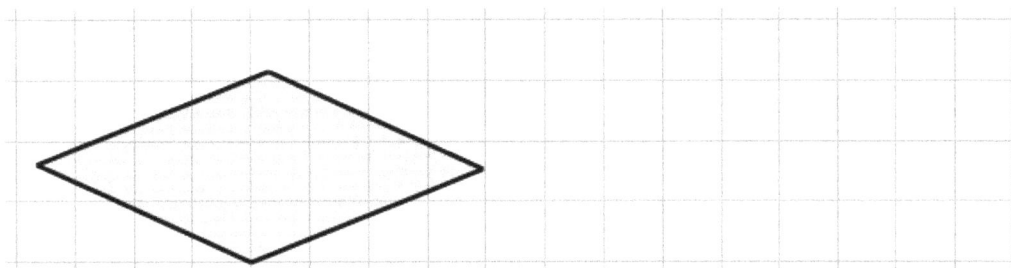

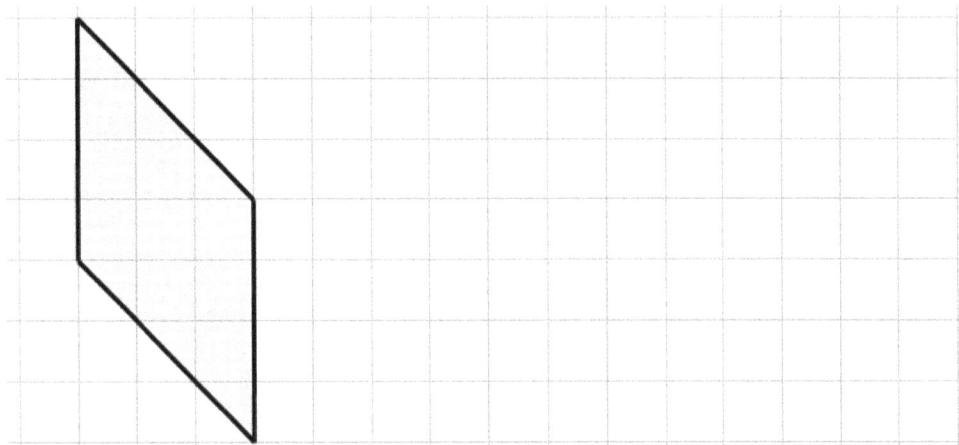

Work out the area of each parallelogram.

Draw a rectangle that you think the parallelogram was taken from.

Now work out the area of the rectangle.

Discuss any differences with your group and compare your answers with theirs.

Discussion

Questions and key points

Complete the following statement: The area of any triangle is _____

that of the area of a rectangle with the same _____ and _____.

Now write a statement similar to the one above, connecting the areas of parallelograms and rectangles.

Where to from here?

What other quadrilaterals do you know? Can you find relationships between these shapes and rectangles with the same dimensions?

MEASUREMENT LESSON

4

Diagonals of Squares: A Dangerous Idea

Introduction

The ancient Greeks thought that all numbers could be written as either whole numbers or fractions. A guy by the name of Hippassus was the first to show this was not true. His "friends" were so unhappy with him that they threw him over the side of the boat they were travelling in.

In this lesson we will explore Hippassus discovery without fear of getting wet.

A.C. Level:	8
A.C. Ref No's:	Investigate the concept of irrational numbers, including π ACMNA186
A.C. Substrands	Real Numbers

Outcome

At the end of this activity:

Students will know:

- The connection between the area of a square and its sides.

Students will be able to:

- The connection between the area of a square and its sides.

Materials Required

Pencil, scissors, calculator.

Activity 1

Tiles and bricks have been used in building, construction and art or thousands of years. They were easier to make than carving stone blocks and could be made to the same shape so that they would fit together neatly without gaps or holes.

Somewhere along the line a tiler with some artistic talent started to draw and paint simple patterns onto plain square tiles. This in turn allowed them, or someone else, to see a something else.

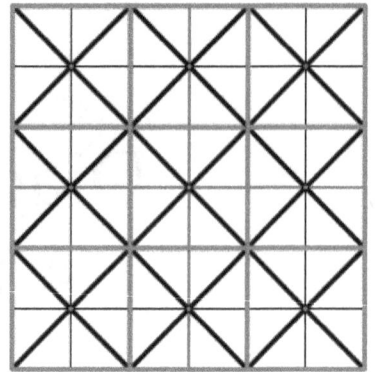

The first part of this activity is to recreate this discovery process. The diagram below is a square with sides of 2 cm in length. Marks have been made at the midpoints of each side.

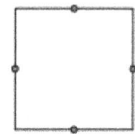

What is the area of this square? _____

Draw lines from top to bottom and from left to right to form four smaller squares.

What is the length of the sides of each of these smaller squares? _____

What is the area of each of these smaller squares? _____

Next, draw diagonal lines cutting each of the smaller squares into triangles with half the area, and that make a square touching the midpoints of the original square. Think carefully about what you have just been asked to do. In order to end up with the correct picture you have to be able to imagine what it looks like and imagination is important for this activity.

What is the area of the square you have just made? If you think it will help you work out the answer, you can copy and cut up the diagram to move the pieces around.

_____

Explain how you know that this is the area.

Can you work out the length of the sides of this square?

If you used a ruler to measure the side lengths you will find that calculating the area using the formula $A = l^2$ gives an answer that is a little bit different to the actual area. This difference is suggesting that there is more to discover than just using a ruler to find length.

To find out what this is examine the set of nested squares in the diagram. Fill in the table to record the side lengths and areas of each one.

	Side Length	Area

For the last few years you have been using the formula $A = l^2$ to find the areas of squares if you are given the lengths. It is possible to use the formula in reverse as well. Sometimes it is easier to "see" what the area of a square is than to measure the lengths of its sides, or its roots like roots of a tree.

The diagram below has the same five squares as before, but with the "roots" of five more growing out of them.

1) Complete the drawings of the five new squares.

2) Work out the area of the new squares.

3) Write the root of the square in the form $\sqrt{A}$.

4) Use a calculator to find the length of each root to five decimal places.

The smallest of the squares has been done for you as an example.

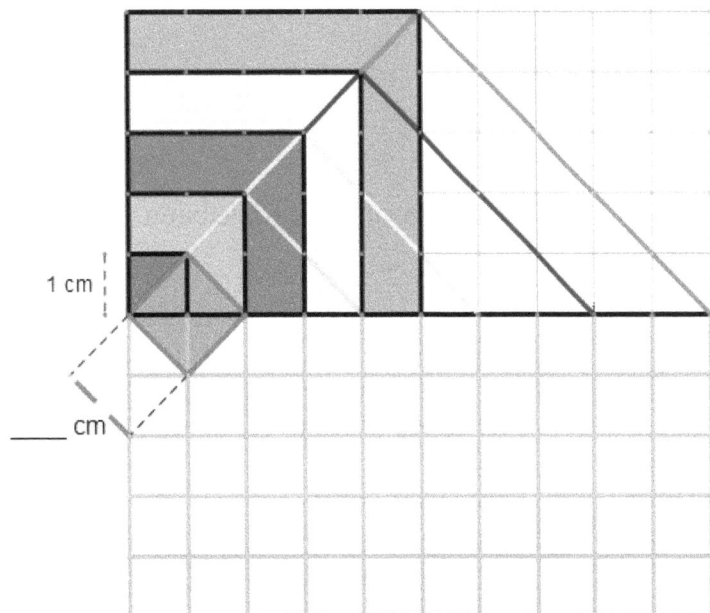

Square Area	Root of the Area	Length of Side
2cm²	√2 cm	1.4142

This method can also be used for finding the roots of squares that are not symmetrical. Here are three more roots. Complete the drawing of the squares that they form and then find the area, root value and approximate length, again to four decimal places.

Square Area	Root of the Square	Approximate side length

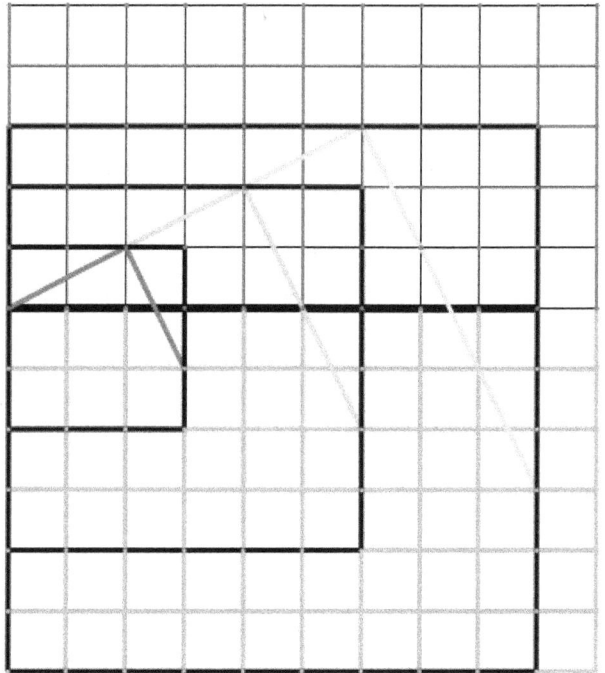

As interesting as this method is, it does not allow us to find and draw the square roots of all whole numbers.

List down the square roots that have been missed so far using this method.

Numbers that are perfect squares, have square roots that can be written as whole numbers or fractions.

For example, 4 is a perfect square and its square root is 2.

2.25 is also a perfect square. It can be written as a fraction, $\frac{9}{4}$. In this form we can take the square root and find it to be $\frac{3}{2}$ or 1.5 as a decimal.

Not all square roots can be expressed like this. This sort of square root is a type of irrational number called a surd. The way to write these numbers is using the square root symbol, for example, the square root of 3 is $\sqrt{3}$. To write a surd in decimal form would be to write down an infinite number of decimal places. For most purposes we just write down the first two or three as an approximation for the exact value.

There are several methods for finding the value of square roots, and for drawing them. A method that lets us "draw" the square root of any whole number is an extension of the one used earlier. As well as drawing the "roots" of a square like before, we also need to draw a semicircle and a diameter.

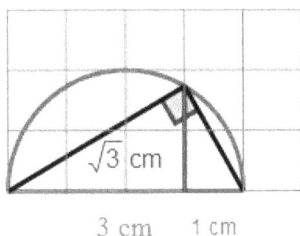

$\sqrt{3}$ cm

3 cm   1 cm

The steps are:

1) Choose a number to find the square root of, for example, 3.

2) Add 1 to get the diameter of the semicircle, D = 3.

3) Use a ruler to draw the diameter and a compass to draw the semicircle, as shown.

4) Divide the diameter into two parts, one part equal to 1 cm and the other equal to the length you want to find the square root of.

5) From the point you have just found, draw a vertical line up to the edge of the semicircle. The resulting length is the square root, in this case $\sqrt{3}$ or approximately 1.73 cm.

Use this method to draw lengths of $\sqrt{2}$ cm, $\sqrt{3}$ cm, $\sqrt{4}$ cm, $\sqrt{5}$ cm and $\sqrt{6}$ cm.

Discussion

Questions and key points

How do we write down the exact value of a square root?

How many decimal places do we usually write down the approximation of a square root? Why do you think this is the case?

Where to from here?

Mathematics in years nine and ten explores the properties of triangles, squares and circles building on from ideas in this activity. The fact that some numbers, actually most numbers, are irrational and cannot be written exactly in fraction form gives us glimpses into the nature of the universe and how it works. Topics with names like Pythagoras' Theorem and Trigonometry form the basis of most building, design and engineering. Without these ideas much of the world around us would not exist.

The number of decimal places that are used in measurements in known as the **precision** of the measurement. Different occupations require different degrees of precision. Find out the level of precision needed for buildings, motor engines, fabrics and materials such as wool, diary (milk) and soft drink manufacture. The last two of these are measuring volume rather than length. Discuss the reasons for these levels of precision.

5

Three Quadrilaterals: Perimeter and Area

Introduction

Quadrilaterals are closed, four sided figures. The areas of all quadrilaterals are connected in various ways to the area of a rectangle of similar size. If we imagine squeezing or stretching these shapes we can also notice some interesting patterns with their perimeters. This lesson starts that investigation process.

A.C. Level:	8
A.C. Ref No's:	Find perimeters and areas of parallelograms, rhombuses and kites ACMMG197
A.C. Substrands	Real Numbers

Outcome

At the end of this activity:

Students will know:

- Both the differences between parallelograms, kites and rhombuses and the similarities.

Students will be able to:

- Calculate the perimeters and areas of parallelograms, kites and rhombuses.

Materials Required

Paper, Pencils, Ruler or Mathomat™

Activity 1

In class you will already have worked with the area formulas for parallelograms, rhombuses and kites. They are given here for you as a reminder.

Parallelogram	Rhombus	Kite
$Area = \dfrac{1}{2} \, base \times height$	$Area = \dfrac{1}{2} \, base \times height$ OR $Area = \dfrac{1}{2} \, x \times y$ where x and y are the diagonals	$Area = \dfrac{1}{2} \, x \times y$ where x and y are the diagonals

Notice that the area of a rhombus can be calculated by either formula, so a rhombus can be thought of as both a parallelogram and a kite.

Similarly, formulas for perimeter for each of these shapes are:

Parallelogram	Rhombus	Kite
$Perimeter = 2B + 2L$	$Perimeter = 4L$	$Perimeter = 2B + 2L$

Label the sides B and L as you see fit. Give an explanation for you placements.

The task for this lesson is to explore what happens to the area and perimeter of each of the three shapes.

Rhombus

Here is a rhombus with side lengths of 6 cm.

First determine both the perimeter and the area of this rhombus.

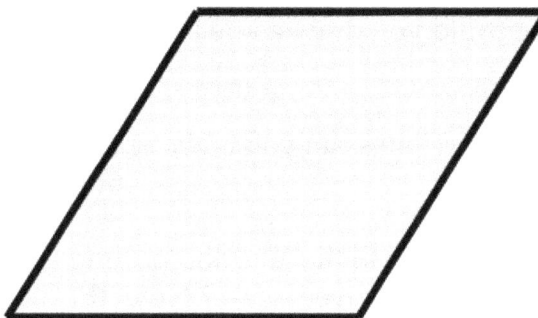

P = _____

A = _____

Now cut the rhombus in half, parallel to the base.

What shape do you have? _____

What is the perimeter of the new shape?

_____

What is the area of the new shape?

_____

Now make another "cut" this time parallel to one of the slanting sides.

What shape do you have? _____

What is the perimeter of the new shape?

_____

What is the area of the new shape?

_____

Now take another rhombus the same size as the original. Find the midpoints of the top and right hand sides and draw a kite as shown.

What is the area of the kite?

A = _____

What is the perimeter of the kite?

P = _____

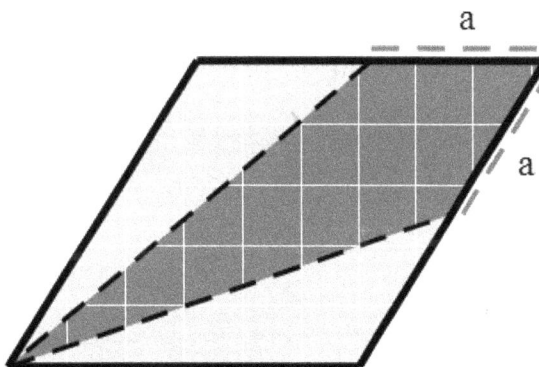

How could you draw a kite inside a rhombus with half the perimeter of the rhombus?

How many kites can you make with perimeters, half that of the original rhombus?

What happens to the area and perimeter of the kite if the length of the short sides, labelled a, are halved again? Note, this will give another kite, with the same long diagonal but a shorter small diagonal.

In the first part of this activity you started with a Rhombus and then made a series of "cuts" to produce parallelograms and kites. In the second part you will be asked to generalise for each of the three shapes. Some of the insights you gained earlier may be helpful but you will need to draw on other ideas as well.

Take each of the three shapes in turn and draw an example. Choose the dimensions carefully to make calculations of area and perimeter as simple as possible.

1) What do you need to do to draw copies of each of these three shapes that have the same size angles but have perimeters half of the originals? Explain all your steps and demonstrate your calculations.

2) What is the area of each of these shapes compared to the originals?

3) What do you need to do to draw copies of the original shapes with half of the area of the originals?

4) What is the perimeter of each of these shapes compared to the originals?

Discussion

Questions and key points

How are rhombuses and parallelograms related? What folds or cuts can you make to turn one into the other?

Similarly, how are kites related to parallelograms and rhombuses?

Where to from here?

There are other types of quadrilateral besides these three. Explore these for connections like the ones found in this activity.

MEASUREMENT LESSON

6

π

Introduction

π is the symbol given to the ratio of the circumference of a circle to its diameter. It is needed if we are to calculate the circumference or area of any circle. It also turns up in many places in nature and mathematics. This lesson explores π and similar ratios for shapes other than circles.

A.C. Level:	8
A.C. Ref No's:	Investigate the concept of irrational numbers, including π ACMNA186
A.C. Substrands	Real Numbers

Outcome

At the end of this activity:

Students will know:

- The origin of the value of π

Students will be able to:

- Measure the value of to successively more precise values.

- Demonstrate that the perimeter and area of polygons gets closer that of a circle as the number of sides of the polygon increases.

Materials Required

Mathomat™, Cash Register Paper rolls, tape measure, circular or cylindrical objects, Dynamic Geometry Software.

Activity

π is the symbol used to represent the ratio between the circumference of a circle and its diameter. In this activity you will firstly explore the value of π for different sized circles and secondly how this ratio can be made for polygons.

The Mathomat has a collection of eight circles of different diameters. Locate the circles with the diameters given in the table. Note that there are also three blank rows in the table. These are for you to add measurements of other circles you can find.

Diameter	Circumference	Ratio $\frac{\text{circumference}}{\text{diameter}}$
15mm		
20mm		
25mm		
30mm		
40mm		
50mm		

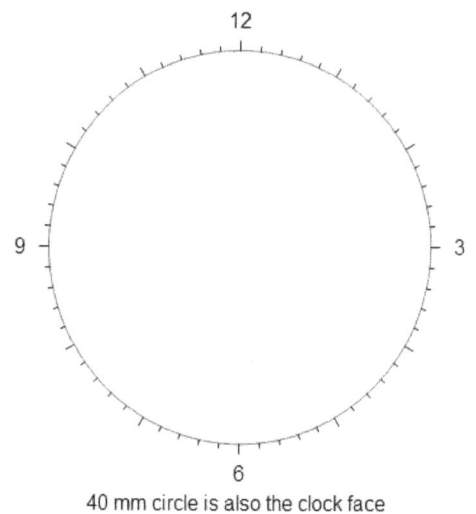

40 mm circle is also the clock face

Next, take a strip of the paper roll and insert it into one of the circular holes in the Mathomat. Allow it to unwrap so that the outside of the roll is in full contact with the circle. You may want to use a book to lie the rest of the Mathomat on for some extra stability.

Use a pen or pencil to mark the circumference of the circle onto your paper strip. You can then take the strip out and lie it flat. Use the mm ruler edge of the Mathomat to measure the length of the circumference for each circle.

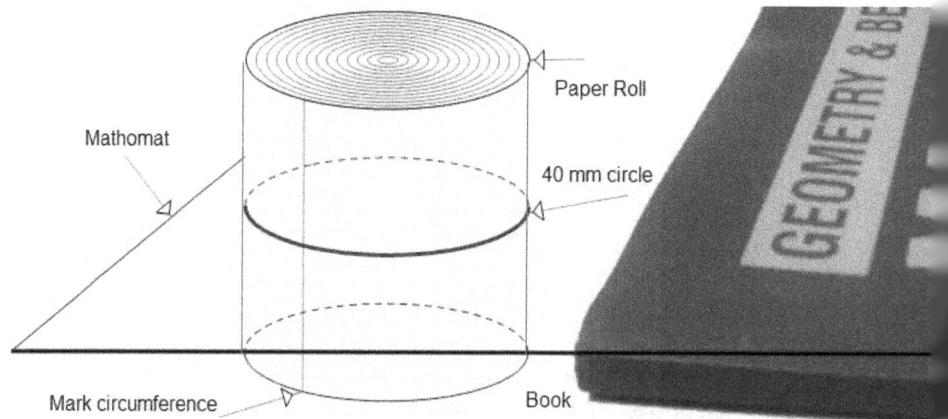

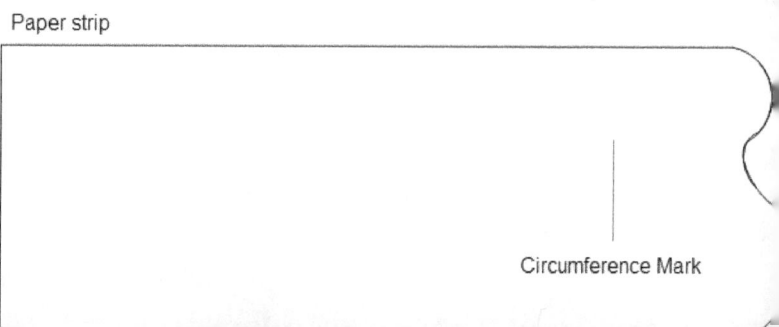

Paper Roll

40 mm circle

Mathomat

Mark circumference

Book

Paper strip

Circumference Mark

Now use your calculator to find the ratio of

$\dfrac{\text{circumference}}{\text{diameter}}$ for each circle.

What do you notice about the calculated values of π for each circle?

_____

_____

Next, look around your classroom and your school to find three very large examples of circular objects. Use the tape measure to find the diameter of each of these and either the tape measure or more paper strip to wrap around the objects to determine their circumferences.

Complete the table and calculate the ratios of $\dfrac{\text{circumference}}{\text{diameter}}$ for each object.

What do you notice about the value of π as your circles become larger?

_____

What length, in millimetres, does the diameter of a circle have to be for the $\dfrac{\text{circumference}}{\text{diameter}}$ ratio to have a calculated value of 3.1416 without rounding? Explain your method for determining this.

Now we will explore the ratio of $\dfrac{\text{circumference}}{\text{diameter}}$ for regular polygons.

The "radius" of a polygon can be taken as the length of the radius of a circle that the polygon can be inscribed in. The "diameter" is then double the radius, just as for a circle. Here are examples of "radius" for a square and a pentagon.

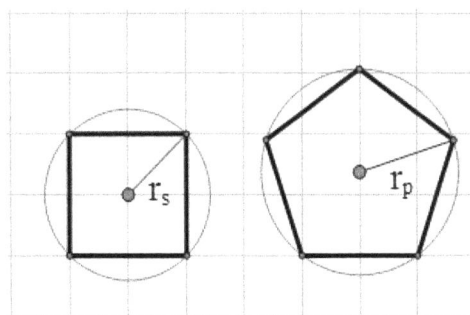

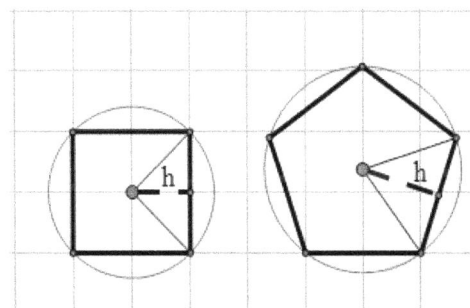

Starting with a side length of 2 cm for each of the polygons, the radius of each of the polygons has been given to four decimal places. You will need to find the perimeter and then calculate the value of $\pi_{perimeter}$ or π_p for short.

The vertical height of the triangles is also given to four decimal places. You can calculate a second value of π, this time π_{Area} or π_A, by rearranging the area formula.

# Sides of Polygon (n)	Radius	Triangle Height	Perimeter	Area	π_p	π_A
3	1.1547 cm	0.5774 cm				
4	1.4142 cm	1.0000 cm				
5	1.7013 cm	1.3764 cm				
6	2.0000 cm	1.7321 cm				
8	2.6131 cm	2.4142 cm				
10	3.2361 cm	3.0777 cm				
12	3.8637 cm	3.7321 cm				
16	5.1258 cm	5.0273 cm				
18	5.7588 cm	5.6713 cm				
20	6.3925 cm	6.3138 cm				
22	7.0267 cm	6.9552 cm				
24	7.6613 cm	7.5958 cm				

Discussion

Questions and key points

1) Do you notice any patterns to the values in your table?

2) What do you predict will happen for very large values of n, that is, for polygons with very many sides?

3) For circles, the value of π used to find both circumference and area is exactly the same. What do you notice about the values of and for polygons?

4) These values of for polygons, while interesting, are not very useful. Why is the real value of π (the one for circles) more significant?

Where to from here?

π is a very special number. It is a basic value, or constant, of the universe. It is an irrational number. There are other numbers that are also constants of the universe. These numbers connect in with the explanations of why the world around us is the way it is. Search out and examine the properties of some of these other constants. Begin with e and ϕ.

SHAPE LESSON

1

Drawing Solids

Introduction

Our world is three dimensional. Every object we see and use in everyday life has height, width and depth. The simplest example of this is a box.

Height

Depth

Width

In books, on TV and computers though, all our shapes are flat, they only have height and width. Exploring the ways in which we represent depth is the purpose of this lesson.

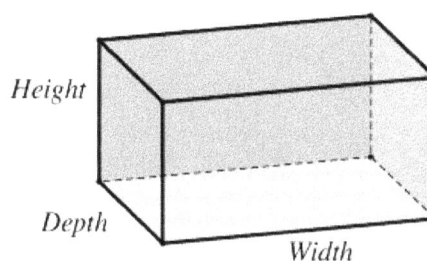

A.C. Level:	5
A.C. Ref No's:	Connect three dimensional objects with their nets and other two dimensional representations ACMMG111
A.C. Substrands	Shape

Outcome

At the end of this activity:

Students will know

- What is meant by the terms Isometric View, Net and Orthographic View

Students will be able to

- Construct nets for a number of different solids

- Draw what solid shapes will look like from different viewpoints using isometric drawing techniques

- Draw what the face views of solids look like using orthographic drawing techniques.

Materials Required

Mathomat™, Netbuilder™, Sketchpad™

Activity 1: Isometric Drawing

This is where solids are drawn as viewed from an angle so that sides in different directions look the same if they have the same length in real life.

Use the 60° rhombus from the prism cluster to make a copy of the cube shown. Each face of the cube is the same size and shape, just rotated about the top front corner.

Solid Cube	See Through Cube

Now use the square, the 60° rhombus and the 30° to make a copy of the face on view of the cube shown.

Solid Cube	See Through Cube

Isometric views of boxes with other dimensions are also able to be drawn using the prism cluster.

Rectangular Prisms: one with a square end and the other with a rectangular end.

What three shapes do you need to draw each of the rectangular prisms?

Triangular prism: Use the Equilateral Triangle for the front and back faces and then join the corners.

Another way to draw Isometric Views of solids is by using Triangular, or Isometric, Grid Paper. On the grid below, the two views of a cube have already been drawn. Use the remaining space to draw the two rectangular prisms and the triangular prism.

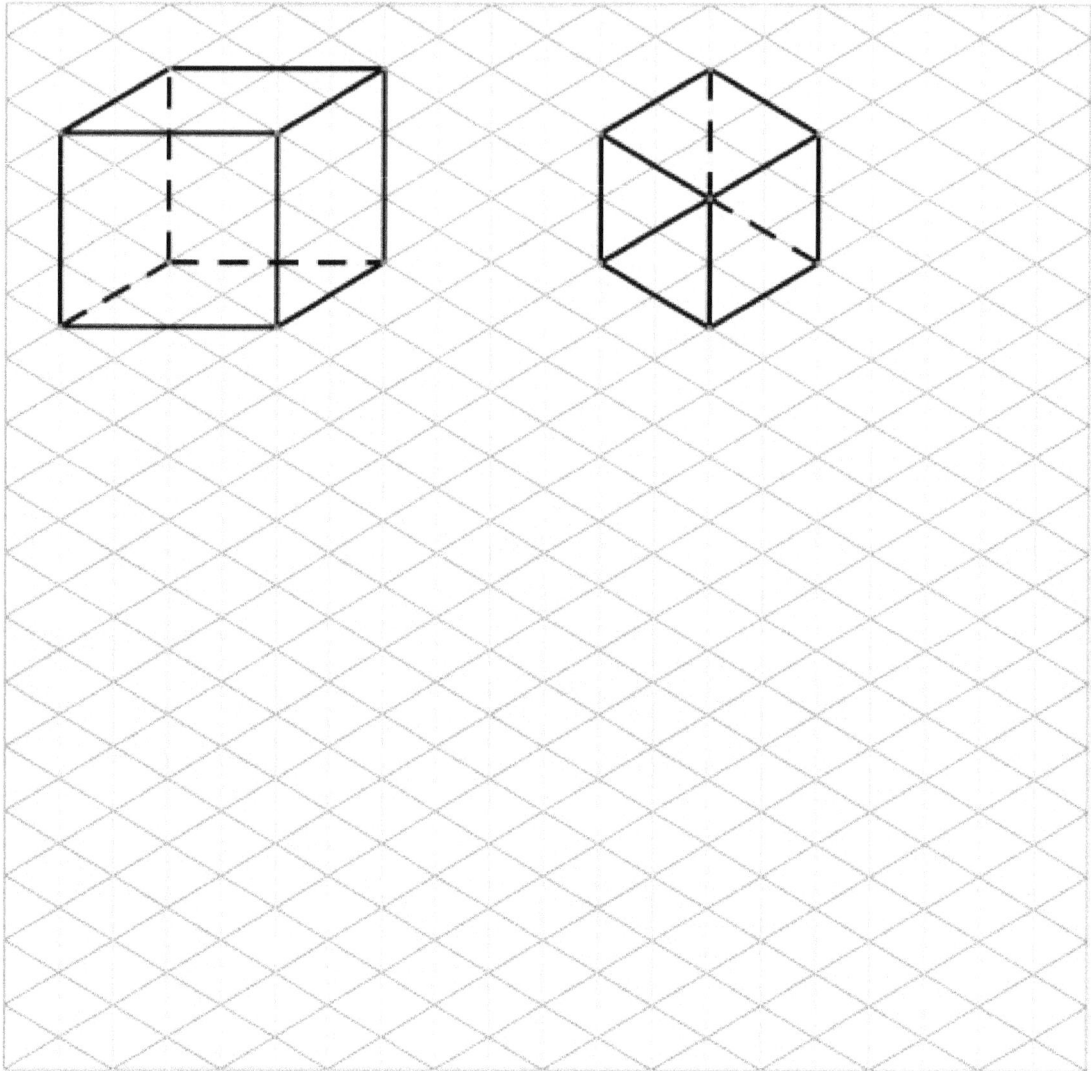

On a new piece of grid paper, now draw:

a Pyramid	Steps

Activity 2: Nets

Solids can be "cut" and opened up and flattened to show what are called net views.

These nets are made up of 2 dimensional shapes: Triangles, Squares, Rectangles and Circles.

Let's start with a cube. Cubes are boxes with six sides. Each side is a square.

An example of a net for a cube is shown here. It uses six squares and six tab ends to allow you to glue or sticky tape the faces together.

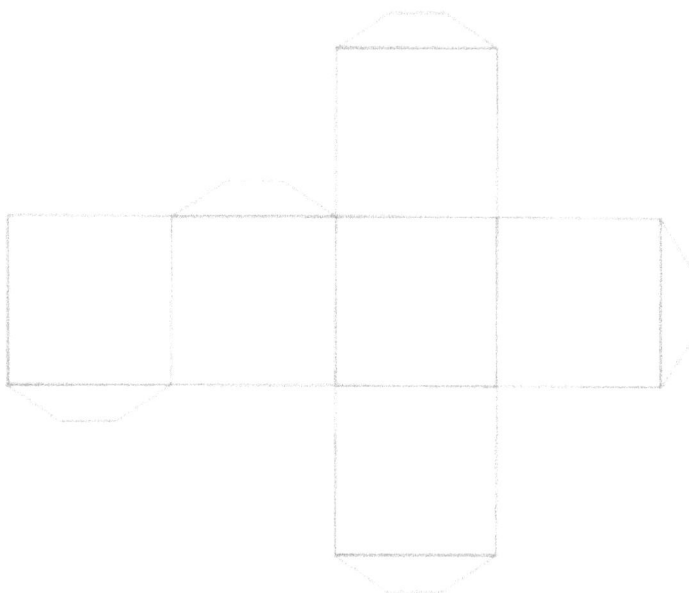

Use your Net Builder to construct your own version of the Cube Net and then cut it out, fold it and stick it together to make the three dimensional shape.

Now use the Net Builder to construct the following solids. Write down the different types of shape you use and how many of each shape.

Solid	Shapes Needed and How Many
Rectangular Prism	
Triangular Prism	
Cylinder	

Activity 3: "House Plan" View[i]

This is where each side of an object and the top and bottom are shown as flat views face on. Consider the following solid and the six views of each of its faces[ii].

Can you spot the differences between:

a)   The back view and the left view?

b)   The front view and the right view?

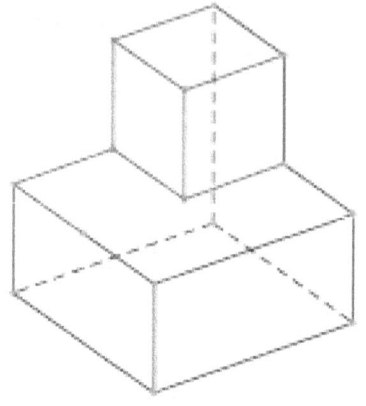

What is unique about the top view of this solid?

Back	Left	Front
Right	**Base**	**Top**

Try drawing your own "House Plan View " for the solid shown.

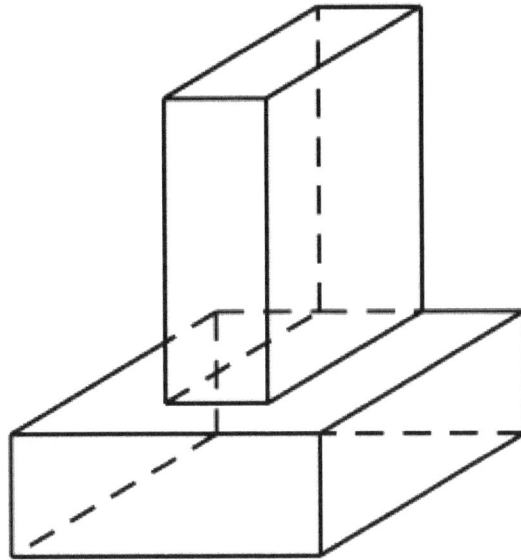

Back	Left	Front
Right	**Base**	**Top**

Now use construction blocks (Lego etc) to build two solids of your own.

Draw them in each of the three different ways you have practiced in this lesson.

Isometric paper is provided, but use blank paper for the nets and house plan views.

Discussion

Questions and key points

Three dimensional objects can be drawn in two dimensions in a number of ways. Which way is easiest for you to "see" the 3 D nature of the original object?

Which way of drawing 3D objects is hardest for your to "see" the nature of the object?

Imagining what objects will look like from different angles and perspectives is a skill required not just in mathematics but many other careers as well. List some of these and explain why the skill is important.

Where to from here?

Developing the ability to imagine objects in your mind and then draw them is valuable. Look around your room, your school, your home. Imagine what some of the objects might look like from different points of view and then use the techniques practiced in this activity to draw them. Then, go and look at the objects from your imagined positions to wee how accurately you were able to get your diagrams.

SHAPE LESSON

2

Prisms and Pyramids

Introduction

AS in Lesson 1 of this chapter, we are again working with drawing two dimensional diagrams of three dimensional objects. This time we are using isometric paper only.

A.C. Level:	6
A.C. Ref No's:	Construct simple prisms and pyramids[iii] ACMMG140
A.C. Sub-strands	Shape

Outcome

At the end of this activity:

Students will know

*   What is meant by the terms Isometric View, Net and Orthographic View

Students will be able to

*   Construct nets for square, rectangular and triangular prisms

*   Construct nets for square based, rectangular based and triangular based pyramids

Materials Required

Mathomat™, Netbuilder™, Sketchpad™

Activity 1:

What is a prism?

Think of a loaf of sliced bread. Each slice looks like all the other slices.

The shape of each slice is the same size and shape as all the other slices.

This is the definition of a prism.

Can you express this in your own words?

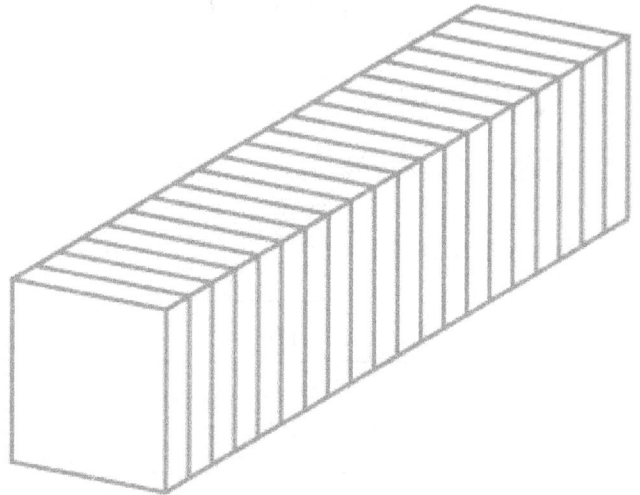

What is a Pyramid?

Here are three different types of pyramid.

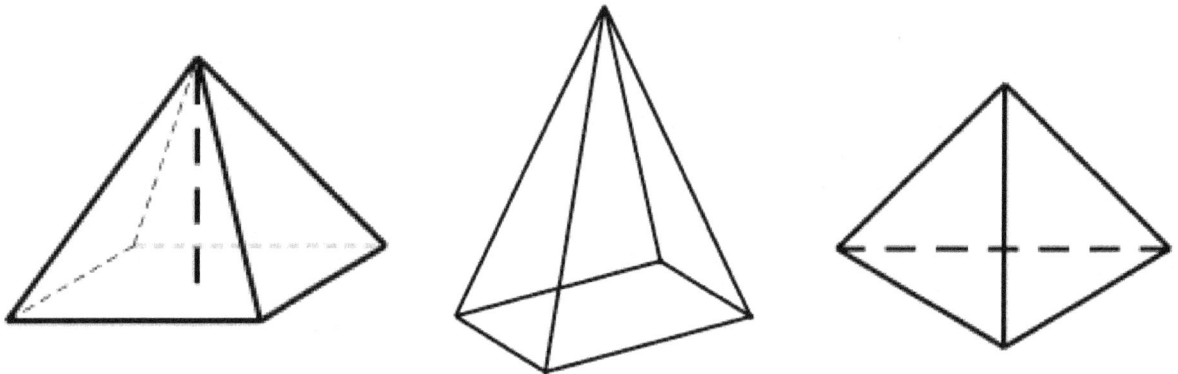

What do they all have in common?

Now look at their bases. Is there a way you could use the shape of each base to describe what type of pyramid is which and how they are different?

For each of the seven solids listed:

1) Draw an isometric view

2) Give each face a name (front, back, left etc)

3) Describe the shape of each face (rectangle, square, equilateral triangle etc)

4) Use your Net Builder to draw a net of the solid

5) Cut out and construct the solid.

Cube

Isometric Drawing	Face Names	Face Shapes

Rectangular Prism

Isometric Drawing	Face Names	Face Shapes

Triangular Prism with Equilateral Ends

Isometric Drawing	Face Names	Face Shapes

Triangular Prism with Right Triangular Ends

Isometric Drawing	Face Names	Face Shapes

Square Based Pyramid

Isometric Drawing	Face Names	Face Shapes

Rectangular Based Pyramid

Isometric Drawing	Face Names	Face Shapes

Triangular Based Pyramid Using Equilateral Triangles for All Faces

Isometric Drawing	Face Names	Face Shapes

Discussion

Questions and key points

Have a look at the way you constructed the nets of your solids. Are they the same or different to the nets made by others in your class.

How many of the solids that you have examined in this activity can be constructed correctly using more than one net?

For each of the prisms, how many different shapes did you have for the faces?

Which shape is most common in nets of the prisms? Explain why these are not used to name the prisms.

Similarly for the pyramids which shape is most common in the nets? Explain why these are not used (in general) to name the pyramids.

Where to from here?

Are there other ways of drawing prisms and pyramids that make more sense to you, that is, help you see and understand what the parts of the solid look like from different points of view? Take the same objects as you have already used and try drawing them using these different styles.

SHAPE LESSON

3

Constructor Challenge[iv]

Introduction

This activity uses the knowledge and skill learned in lessons 1 and 2 of the Shape chapter and asks you to solve two problems while working in teams. Good luck.

A.C. Level:	7
A.C. Ref No's:	Draw different views of prisms and solids formed from combinations of prisms ACMMG161
A.C. Sub-strands	Shape

Outcome

At the end of this activity:

Students will know

• How to construct and draw solids of various designs.

Students will be able to

• Construct composite solids from simpler prisms and pyramids

• Draw what these composite shapes will look like from different angles

Materials Required

Mathomat™, Netbuilder™, Construction Blocks (Eg: Lego™), Sketchpad™

Activity 1:

Using either the solids made from activity 2 or construction blocks students are to complete two team challenges.

Challenge 1

Design, construct and then draw isometric views of sophisticated three dimensional objects.

Challenge 2

Take photographs of their objects from various angles.

Choose one photograph to show to other teams.

Opposing teams then have to draw the object from at least one other angle using just the photograph they have been given.

Discussion

Questions and key points

For challenge 1, what component objects did you use to make your model?

For challenge 2, why did your team choose the photo you gave to the other teams? What were the elements that you thought would be challenging to them?

Again for challenge 2, what did you find challenging about the photo(s) given to your team?

How successful was your team at drawing alternate views from the photos you were given?

Where to from here?

In what careers is the ability to visualise and draw views of objects from different perspectives useful? Find out more about these careers.

---

i Teacher Note: The technical term for this is Orthographic View but it would only be worth mentioning to students at year 9 or 10 level.

ii It may be useful for teachers to instruct students to colour in the different faces, and parts of faces, in different colours.

iii This A.C reference assumes students are already familiar with the square, rectangle, and different types of triangle. If this is not the case then introductory work needs to be done in developing this knowledge.

iv This third lesson follows on from the first two in conceptual development and difficulty. Even though the three tasks are separated by a year level each according to the Curriculum there is benefit in conducting at least Lessons Two and Three together in year 7.

LOCATION AND TRANSFORMATION LESSON

1

Map Grids

Introduction:

Location is about establishing one's place in space and being able to move around in a repeatable manner. In order to do this we need to develop a sense of where we are and connect this with other places. At the very least we need to know where to start, where to finish and have a method for describing how to move from one to the other.

A.C. Level:	5
A.C. Ref No's:	Use a grid reference system to describe locations. Describe routes using landmarks and directional language ACMMG113
A.C. Substrands	Location and Transformation

Outcome

At the end of this activity:

Students will know:

- How to use grid reference systems

Students will be able to:

- Find and describe the position of various notable buidlings on maps

- Determine the distances in real life between pairs of locations on maps.

Materials Required

Mathomat™ – 1:20 000 scale rule (This is 1 cm = 200 m)

Using Google Maps™ or Whereis™ zoom in until you have the scale showing 200 m.

Copy the view and past it into a Word document.

Next use the word processor ruler to match this with 1 cm.

Activity 1

1) Begin with a map of your town or suburb, with the school clearly visible. Maps are drawn with North at the top and East, South and West at the other three corners.

2) Draw the path from the school to your home (or as far as the map will let you).

3) Describe in words how to walk, ride or drive this route.

4) On another copy of the map, have your partner draw the same path according to your spoken instructions without letting them look at your original.

5) Compare your path with theirs. Talk about and then write down any differences between the two paths.

6) Swap roles and repeat the process.

7) Now use the 1: 20 000 scale on your Mathomat to measure how far you walk or drive from your school to your home or to the edge of the map.

Discussion

Questions and key points

How much effort did it take to make the picture of the map size up correctly so that 1 cm on the map match up with 200 m marker?

Why is it important to make sure you have your scales correctly matched?

Explain why it is usually not possible to travel in a straight line between two places.

Where to from here?

Surveyors and cartographers are responsible for making maps.
Find out more about these two careers and the skills that they need.

Geometry & Beyond with Mathomat

LOCATION AND TRANSFORMATION LESSON

2

Wallpaper Patterns

Introduction:

There are only seventeen different ways to produce patterns for wallpaper. In this lesson you will explore the five simplest patterns and then use what you have learned to make your own designs.

A.C. Level:	5
A.C. Ref No's:	Describe translations, reflections and rotations of two-dimensional shapes. Identify line and rotational symmetries ACMMG114
A.C. Substrands	Location and Transformation

Outcome

At the end of this activity:

Students will know

- The terms for describing reflective and rotational symmetries.

Students will be able to

- Create and describe five different tessellating patterns.

Materials Required

Mathomat™, Hand Mirror, Grid Paper

Part 1

1) Find the 25 mm square, labelled number 1, in the top left corner of your Mathomat. Use this to draw a square on a blank sheet of paper.

2) Place the hand mirror over the square so that it covers exactly half of it, as in the photograph.

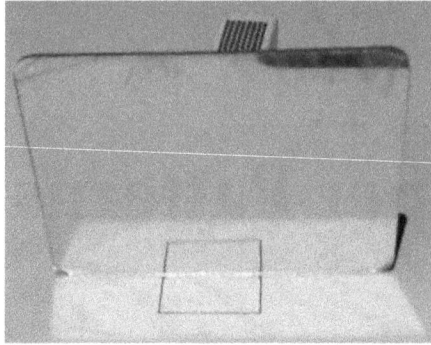

3) Find a position to look at the square so that with or without the mirror it looks exactly the same size and shape.

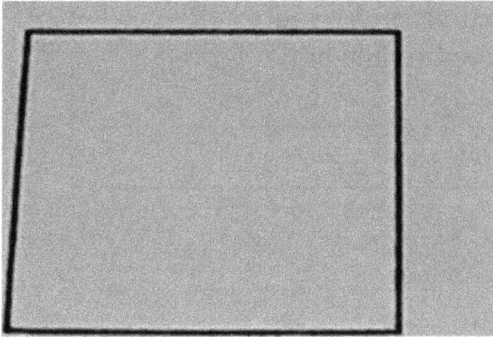

Original Square

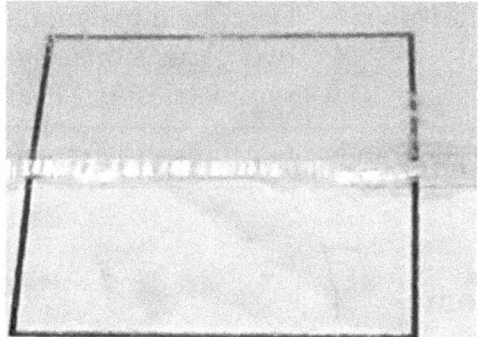

Square and Mirror

4) The line made by the edge of the mirror on the paper is called the line or axis of reflection. Mark in this line on your diagram.

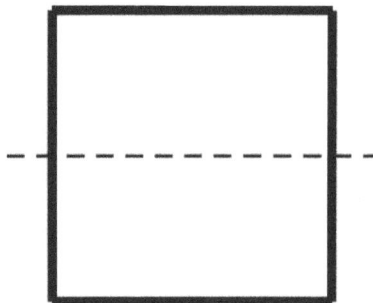

5) Now, to avoid confusion, label the corners of the square with the letters A, B, C and D as shown.

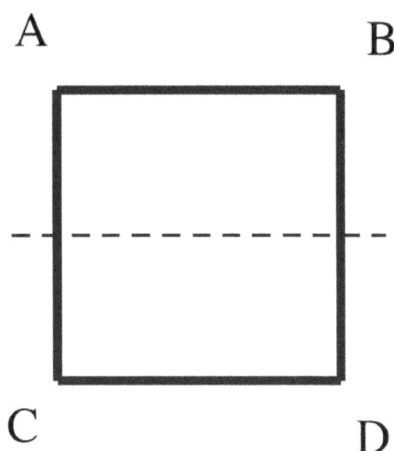

A   B

C   D

6) There are other lines that can be drawn that also act as axes of reflection or axes of symmetry. Use your ruler and the mirror to fins each of them and draw them on to your square.

7) Now draw up a table, with the following shapes in the first column and the number of axes of symmetry for the shape in the last column.

Shape	Mathomat Figure Number	Number of Axes of Reflection
25 mm Square	1	
Rectangle	11	
Parallelogram	7	
Rhombus	12, 16 or 19	
Equilateral Triangle	23	
Scalene Triangle	27	
Isosceles Triangle	33	
Trapezium	24	
Pentagon	25	
Ellipse	4	
Circle	2	

Part 2

1) Start again with the 25 mm square and a blank piece of paper. Draw two copies of the square, with a gap between then

2) Find the centres of both squares by drawing in the two diagonals.

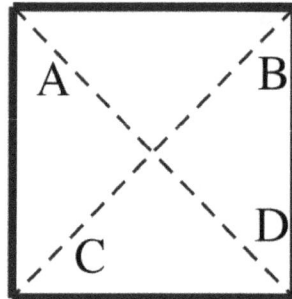

3) Cut out one of the squares and then place it on top of the other one so that they match up perfectly.

4) Place a pen or pencil on the centre of the squares to hold the top one in place.

5) Rotate or turn the top square until it again fits completely on top of the square on the bottom. What you have just found is that squares have Rotational Symmetry just like they have Reflective Symmetry.

6) How many times in a single rotation can a square be rotated so that it looks the same as it does at the beginning? Count a full rotation as one of these.

7) Fill in the table below for the square and then all of the other shapes listed.

Shape	Mathomat Figure Number	Number of Axes of Reflection
25 mm Square	1	
Rectangle	11	
Parallelogram	7	
Rhombus	12, 16 or 19	
Equilateral Triangle	23	
Scalene Triangle	27	
Isosceles Triangle	33	
Trapezium	24	
Pentagon	25	
Ellipse	4	
Circle	2	

Wallpaper Patterns

P1

Simple cell

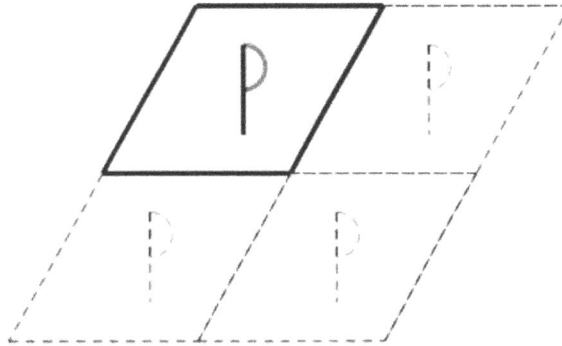

Plane is filled just by translating the original

P2

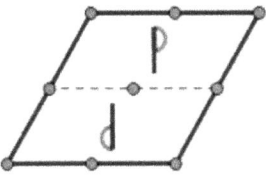

This cell has two components

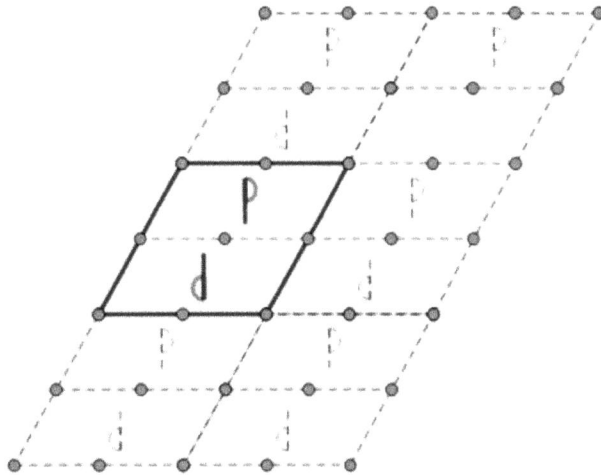

Plane is filled just by rotating the original cell by 180° around any of the red dots

Cell shape for P1 and P2 is a parallelogram.

PM

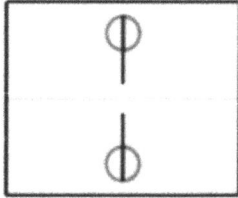

The starting cell has two parts, one the reflection of the other.

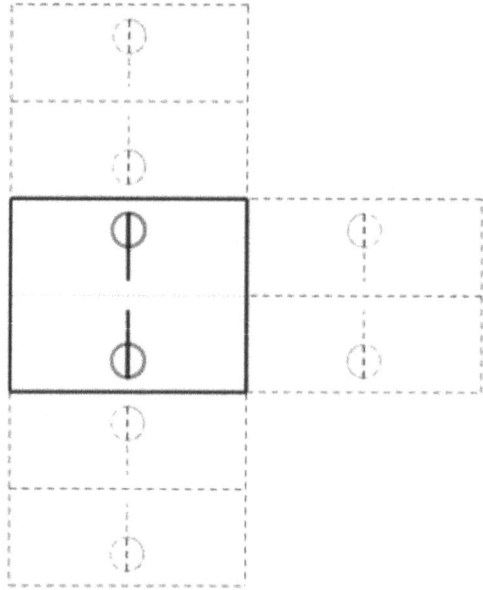

Plane is filled by reflecting in axes. (In this case these correspond to the sides of the orignial but they do not have to.

PG

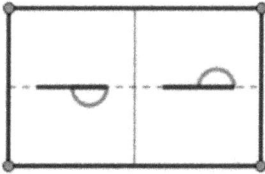

Original Cell has two parts. The right section is the left shifted or translated and then reflected in the dotted axis.

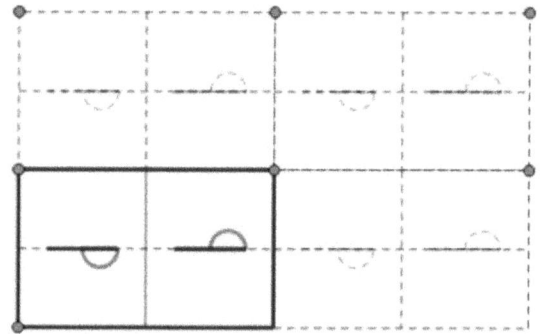

The cell shapes for PM and PG are rectangular.

P4

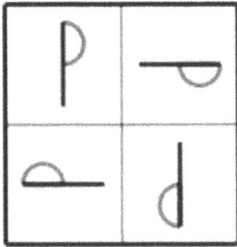

The original cell has 4 parts, all rotations of each other by 90° about the centre.

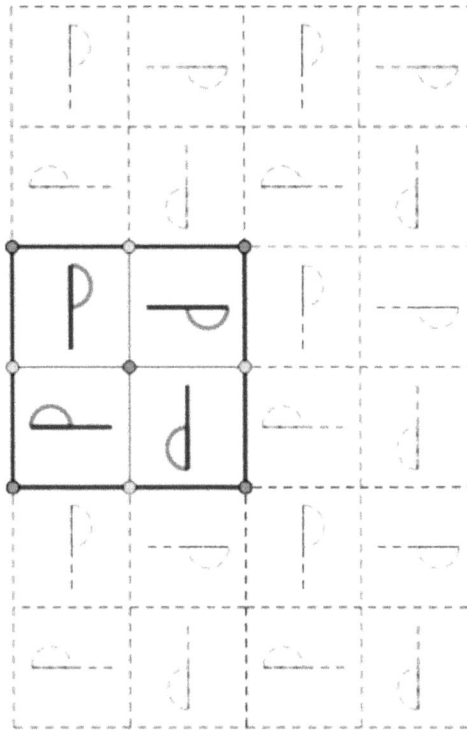

The plane can be filled by rotating by 90° around the red dots or by 180° about the blue dots

The cell shape for P4 is square.

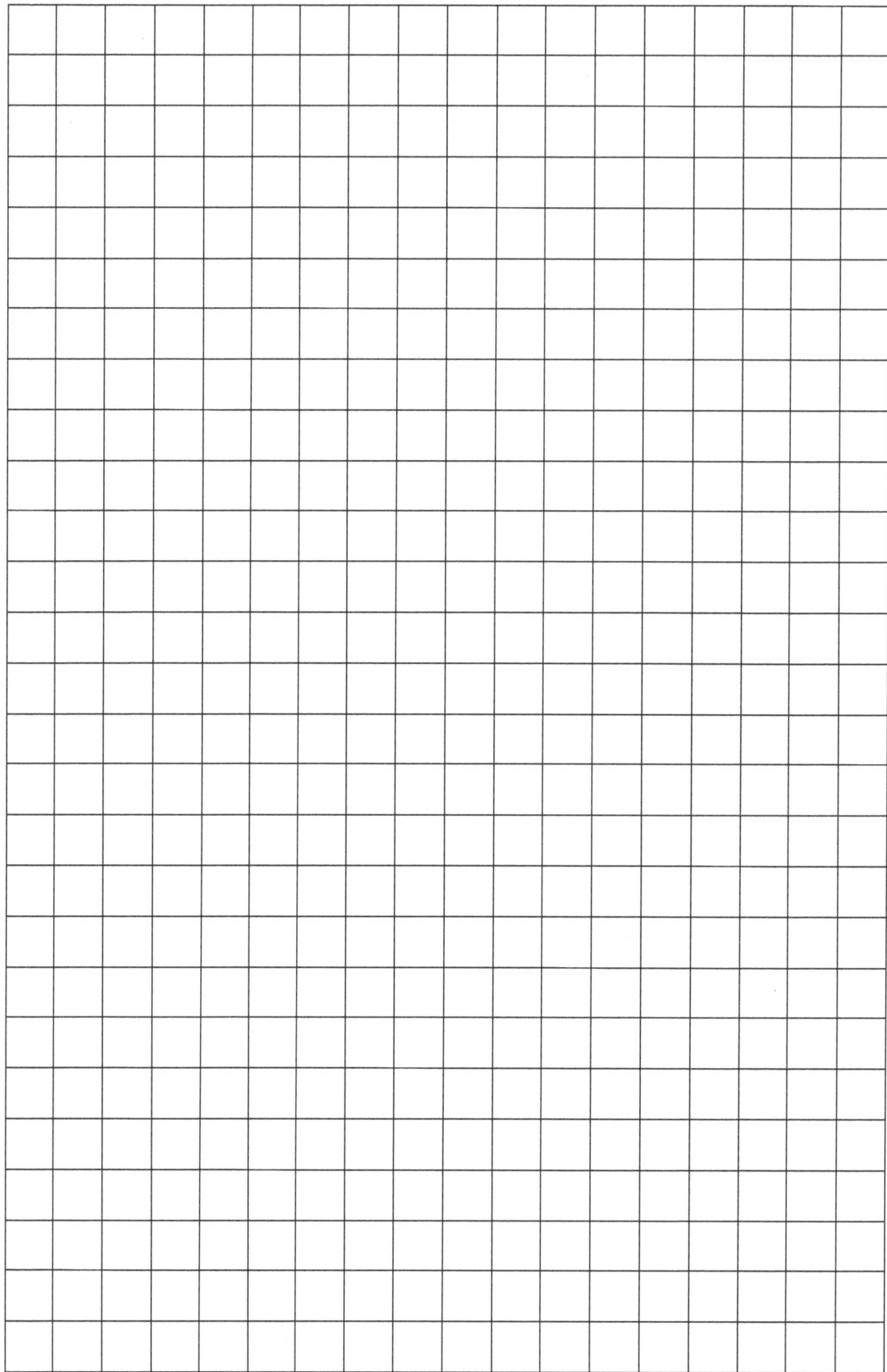

Discussion

Questions and key points

Are there any of the shapes that do not have axes of symmetry? Which ones?

In discussion with a partner, or the class, come to an explanation that you agree on for why this happens.

What is special or unique about the circle? Why?

Where to from here?

As mentioned in the introduction to this lesson, there are a total of seventeen different wallpaper patterns. You might like to investigate some the other twelve in a similar fashion.

PYTHAGORAS AND TRIGONOMETRY LESSON

1

Right Angled Triangles are Special

Introduction:

The sides of right angled triangles are connected by a rule that does not work in any other type of triangle. This lesson explores the nature of this special property.

A.C. Level:	9
A.C. Ref No's:	Investigate Pythagoras' Theorem and its application to solving simple problems involving right angled triangles
A.C. Substrands	Pythagoras and Trigonometry

Outcome

At the end of the activity students will know,

- That Pythagoras Theorem only works for right angled triangles

- The formula for Pythagoras Theorem

- That the results of applying the formula can be integers, fractions or irrational numbers

Students will be able to

- Students will be able to determine the values for unknown lengths in right angled triangles.

Materials Required

Paper, Compass, Ruler.

Activity 1

On the next two pages are two sets of triangles. The first set of triangles all have one side of length three and another of length four. The third side in each triangle is what is different.

1) Find the length of each side according to the dot marks along it. Write down the lengths of all three sides for each triangle.

2) Draw squares on the outside of each triangle based on the length of each of the sides. See the diagrams on the side of this page to see how.

3) Calculate the area of each of the squares and write the answers inside the square

4) Use the answers to complete the following statement:

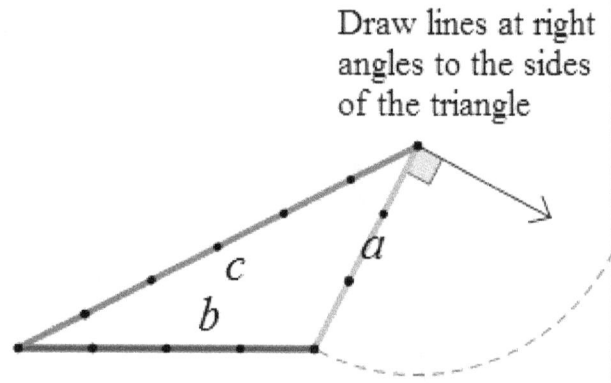

For each triangle cross out the two symbols that are not true out of

    greater than >
    equals =
    less than <

5) Write down what you notice about the three different cases in each set.

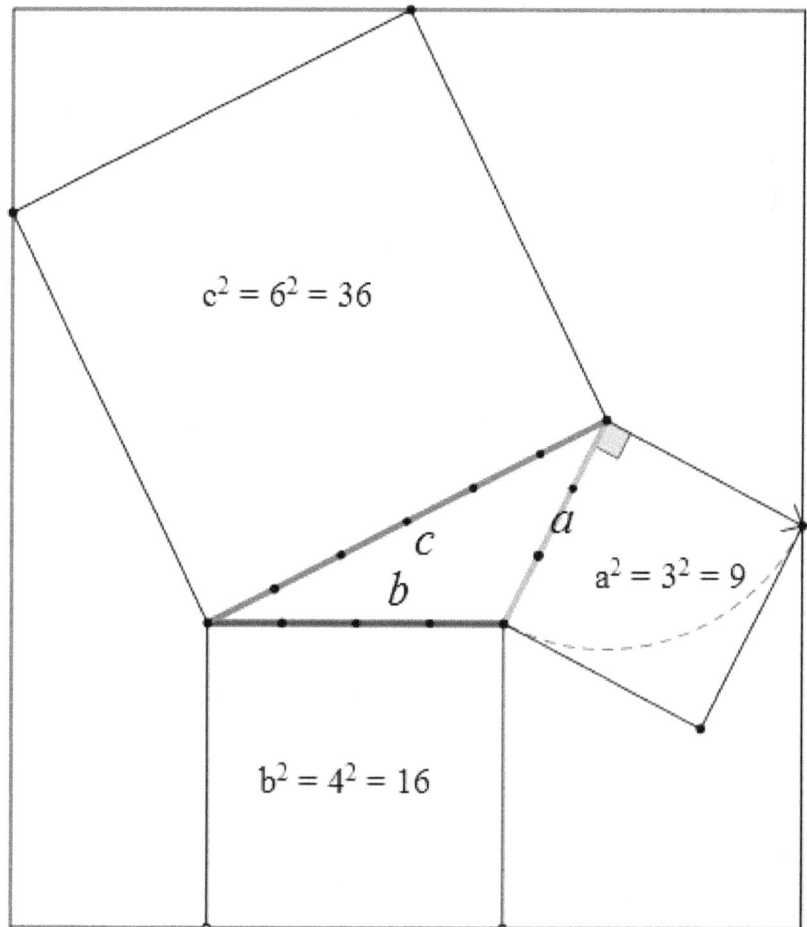

Draw lines at right angles to the sides of the triangle

Use a compass to mark off the length of the sides.

c
b
a

$c^2 = 6^2 = 36$

c
b
a

$a^2 = 3^2 = 9$

$b^2 = 4^2 = 16$

Side lengths:

a =

b =

c =

$$\begin{matrix} > \\ a^2 + b^2 = c^2 \\ < \end{matrix}$$

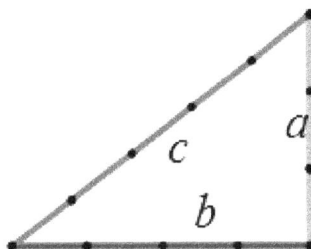

Side lengths:

a =

b =

c =

$$\begin{matrix} > \\ a^2 + b^2 = c^2 \\ < \end{matrix}$$

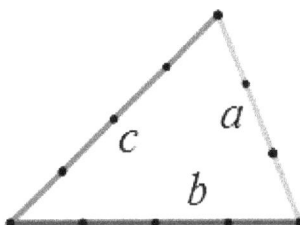

Side lengths:

a =

b =

c =

$$\begin{matrix} > \\ a^2 + b^2 = c^2 \\ < \end{matrix}$$

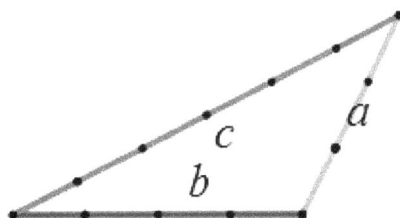

Conclusions:

Side lengths:

a =

b =

c =

$$\begin{array}{c} > \\ a^2 + b^2 = c^2 \\ < \end{array}$$

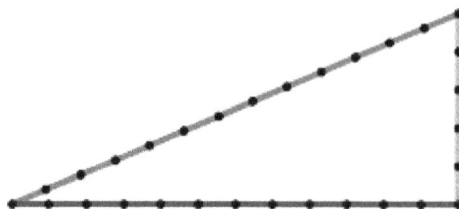

Side lengths:

a =

b =

c =

$$\begin{array}{c} > \\ a^2 + b^2 = c^2 \\ < \end{array}$$

Side lengths:

a =

b =

c =

$$\begin{array}{c} > \\ a^2 + b^2 = c^2 \\ < \end{array}$$

Conclusions:

In which of the cases for each set of triangles was $a^2 + b^2$ equal to c^2?

What type of triangle is this statement true for?
Hint: measure the angles of the triangles.

What you have just discovered is called Pythagoras' Theorem?

Now make your own right angled triangles on dot paper, and draw the squares on the sides, as before. Then, working with your neighbours, fill out the table at right. Use a combination of whole numbered and non-whole numbered sides.

(Note: The "small" and "medium" squares can be the same size.)

The sides of the first four triangles to use have been given to you.

Area of Squares			
Sides	Small	Medium	Large
5,12,13			
6,8,10			
4, 7.5, 8.5			
5, 7,			

3. Describe the pattern of the numbers in the table. Does it always obey the formula?

4. In your own words, state the Pythagorean theorem by completing this sentence: "In a right triangle . . . "

Discussion and wrap-up

Questions and key points

What have you discovered about right angled triangles?

What have you discovered about acute angled triangles?

What have you discovered about obtuse angled triangles?

Where to from here?

Pythagoras' Theorem is used in building and construction industries on a daily basis.   Identify ten different situations or parts of buildings where Pythagoras' would have been used.

PYTHAGORAS AND TRIGONOMETRY LESSON

2

Taxi-Cab Distances

Introduction:

The shortest distance between two points is a straight line, however it is not always possible to travel this way. This lesson looks at what happens to distance when travel is constrained to grids like taxi-cabs on roads.

A.C. Level:	9
A.C. Ref No's:	Find the distance between two points located on a Cartesian plane using graphical and algebraic techniques, including graphing software. ACMNA214
A.C. Substrands	Linear and non-linear relationships

Outcome

At the end of the activity students will know,

* How Pythagoras' Theorem relates to distance.

Students will be able to

* Use Pythagoras' Theorem to find the distance between two points on the Cartesian Plane

Materials Required

Graph or dot paper

Taxicab Versus Euclidean Distance

In towns and cities in order for people to travel from one place to another we have to follow the streets. For the most part streets are set out in a grid where they are either parallel or perpendicular to each other. In this activity we will call this type of travel taxicab travel and the distance covered to get from place to place is the taxicab distance.

If you can travel only horizontally or vertically (like a taxicab in a city where all streets run North-South and East-West), the distance you have to travel to get from the origin to the point (2, 3) is 5. This is called the taxicab distance between (0, 0) and (2, 3). If, on the other hand, you can go from the origin to (2, 3) in a straight line, the distance you travel is called the Euclidean distance, or just the distance.

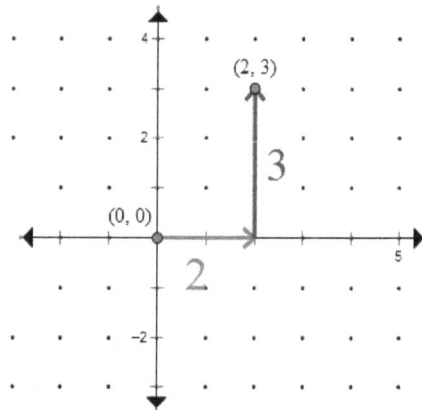

Finding taxicab distance:

Taxicab distance can be measured between any two points, whether on a street or not. For example, the taxicab distance from (1.2, 3.4) to (9.9, 9.9) is the sum of 8.7 (the horizontal component) and 6.5 (the vertical component), for a total of 15.2.

1. What is the taxicab distance from (2, 3) to the following points?

a. (7, 9)

b. (- 3, 8)

c. (2, -1)

d. (6, 5.4)

e. (-1.24, 3)

f. (-1.24, 5.4)

Finding Euclidean distance:

There are various ways to calculate Euclidean distance. Here is one method that is Pythagoras' Theorem.

Since the area of the square at right is 13, the side of the square-and therefore the Euclidean distance from, say, the origin to the point (2,3)-must be , or approximately 3.606 units.

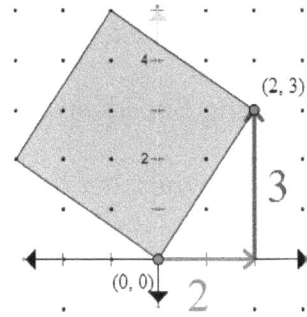

Explain why this is true.

2. What is the Euclidean distance from (2, 3) to the following points?

a. (9,7)

b. (4, 8)

c. (2, 5.5)

d. (6, 0)

e. (1.1, 3)

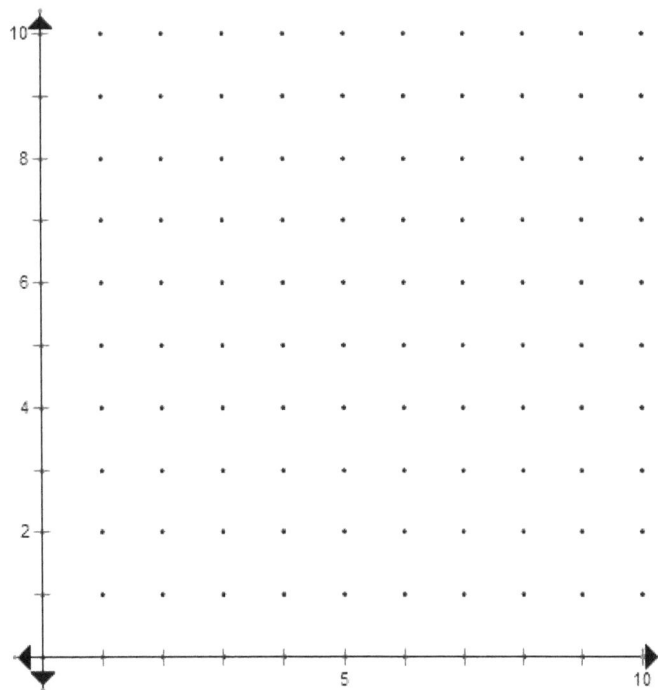

3. Find as many points as possible on the grid below that are at a distance 5 from (5, 5).

a. Using taxicab distance

b. Using Euclidean distance

c. Using the ruler on your Mathomat.

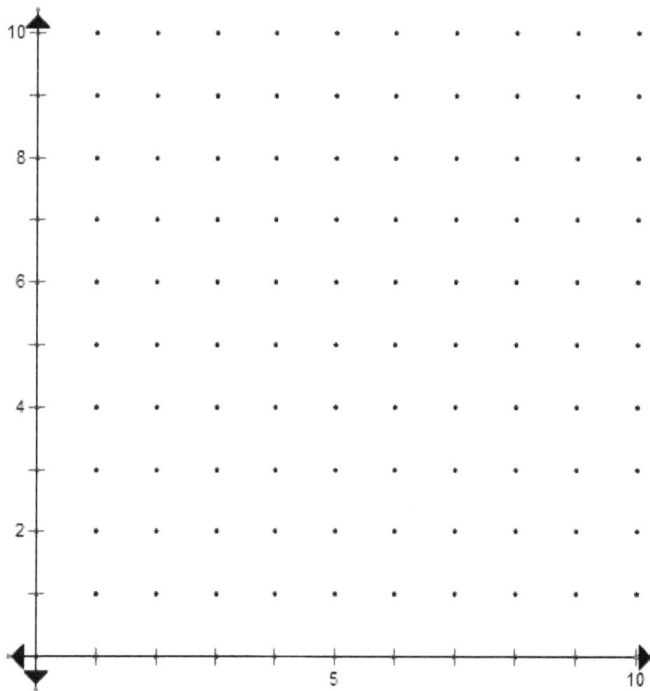

Discussion

Questions and key points

Find a formula for the taxicab distance between two points P ($x1$, $y1$) and P2 ($x2$, $y2$). Call the distance T(P1, P2)
(Hint: Start by figuring out a formula, for the case where the points are on a common horizontal or vertical line. The formula should work whether P1 or P2 is named first.)

In Euclidean geometry, for three points A, B, and C, we always have
AB + BC > AC

This is called the triangle inequality. Does this work in taxicab geometry? In other words, do we have T(A, B) + T(B, C) > T(A, C)? If so, in what cases do we have equality?

Which is usually greater, taxicab or Euclidean distance? Can they be equal? If so, in what cases?

Explain why the answers to Problem 3a are located on what may be called a taxi-circle.

Where to from here?

There are many cases in real life where it is not possible or practical go take a straight line path between two points. Make a list of some of these. As well as the Pythagoras Theorem, what other ideas in mathematics are used to work out answers in these situations?

PYTHAGORAS AND TRIGONOMETRY LESSON

3

Right Angled Ratios

Introduction:

If two right angled triangles each have an angle with the same value then the triangles are related to each other. This means that we can use what we know about the sides and angles of one triangle to find out information about the other one. This lesson introduces the concepts behind this relationship.

A.C. Level:	9
A.C. Ref No's:	Use similarity to investigate the constancy of the sine, cosine and tangent ratios for a given angle in right-angled triangles ACMMG223
A.C. Substrands	Pythagoras and Trigonometry

Outcome

At the end of the activity students will know,

- That pairs of sides in right angled triangles have constant ratios for any given angle, no matter how big the triangle.

Students will be able to:

- Calculate values for the three ratios known as sine, cosine and tangent

Materials Required

Mathomat, Radial Grid Paper

Radial Grid Paper

This type of grid paper is made up of two sets of measurements. The first set forms a series of arcs or circles with radii marked in one unit steps. This allows us to find the length of a diagonal line. All lines of the same length drawn from the bottom left corner will end on the same arc.

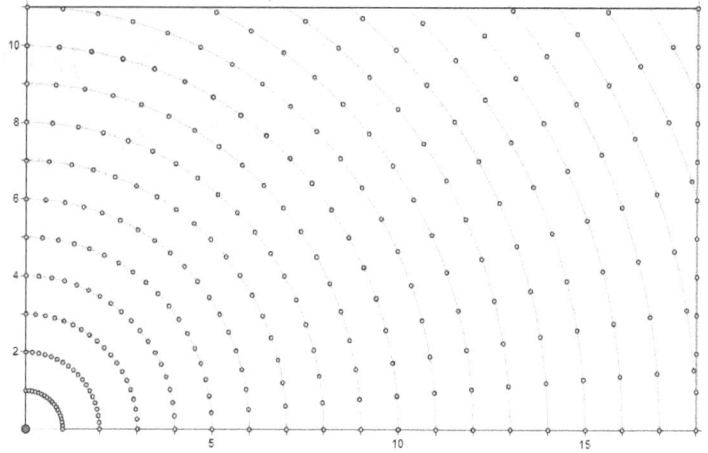

Draw diagonal lines starting at the origin that are length :

a) 2   b) 4   c) 6   d) 8   e) 10 units

(Note we are not measuring in cm, but in the size of a unit on this paper.)

The second set of measurements are angles. The angles are measured anti-clockwise from the horizontal. To help keep track of the size of the angles some guidelines have been added in at 15°, 30°, 45°, 60°, 75° and 90°.

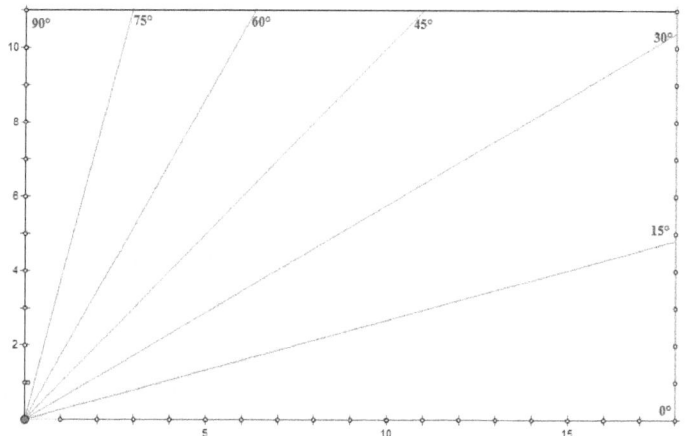

Using the sample grid here, use your protractor to find and mark in angles of:

a) 10°   b) 20°   c) 40°   d) 50°   e) 70°   f) 80°

Now consider the diagram below. Onto the Radial Grid Paper has been drawn a number of right angled triangles with a common angle of 30°.

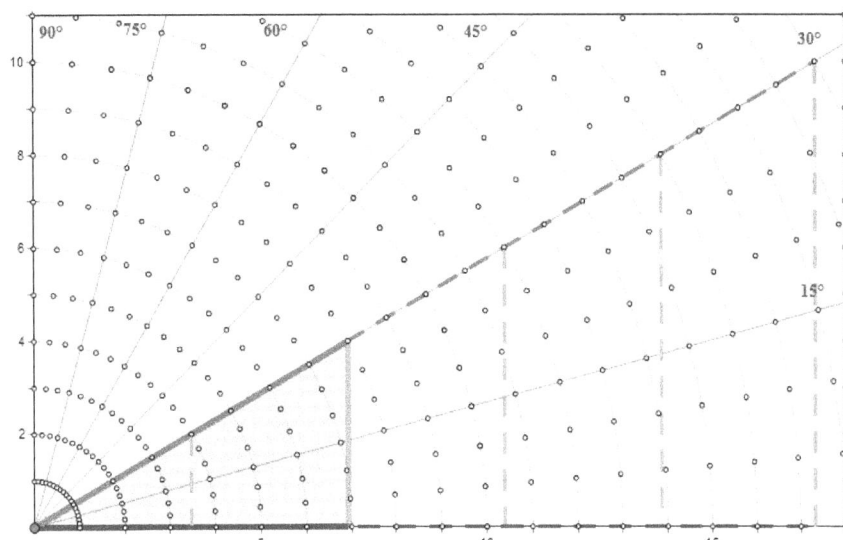

These triangles are said to be similar to one another. This can be checked by using the Angle-Angle-Angle Rule for similarity. Work with one or two other members of your class to write a clear statement of this for the triangles shown.

Using the grid paper references, fill in the table of values for the Opposite sides of each of the triangles. You will find they are always a whole number value.

Opposite					
Adjacent					
Hypotenuse	2	4	6	8	10

Now use Pythagoras' Theorem to calculate the length of each of the Adjacent sides. Write your answers to three decimal places.

Using the values in your table, now construct a second table by calculating the ratios of pairs of sides of the triangles.

$\frac{Opposite}{Hypotenuse}$					
$\frac{Adjacent}{Hypotenuse}$					
$\frac{Opposite}{Hypotenuse}$					

Now consider the diagram below. Onto the Radial Grid Paper has been drawn one of a number of right angled triangles with a common angle of 45°.

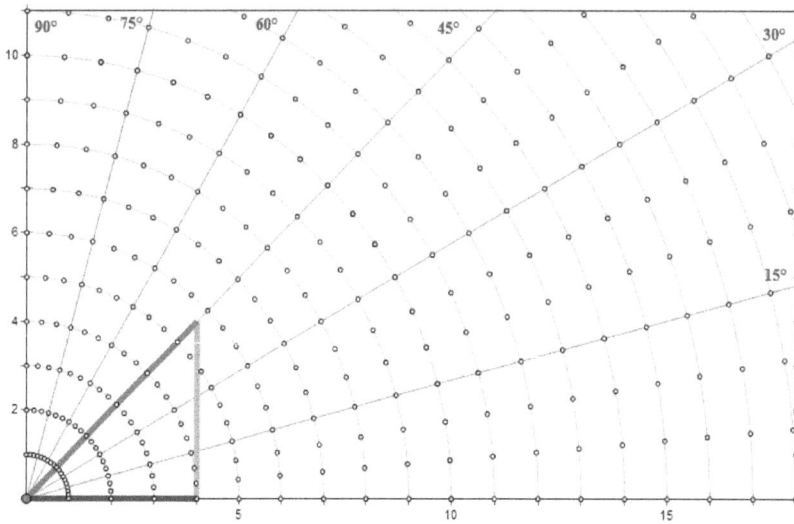

Using the grid paper references, draw the other four triangles and fill in the table of values for the Adjacent sides of each of the triangles. You will find they are always a whole number value.

Opposite	2	4	6	8	10
Adjacent					
Hypotenuse					

Now use Pythagoras' Theorem to calculate the length of each of the Hypotenuses. Write your answers to three decimal places.

Using the values in your table, now construct a second table by calculating the ratios of pairs of sides of the triangles.

$\frac{\text{Opposite}}{\text{Hypotenuse}}$					
$\frac{\text{Adjacent}}{\text{Hypotenuse}}$					
$\frac{\text{Opposite}}{\text{Hypotenuse}}$					

Now consider the diagram below. Onto the Radial Grid Paper has been drawn the first of a number of right angled triangles with a common angle of 60°.

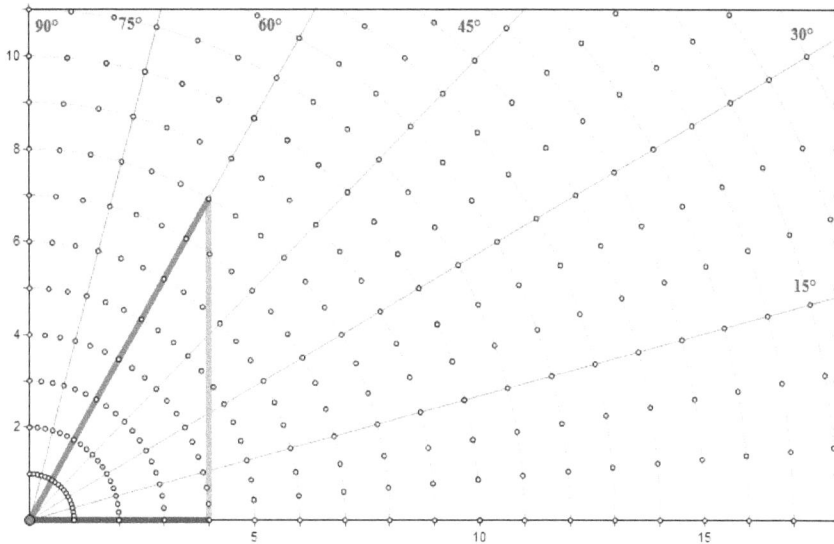

Using the grid paper references, draw in the other four triangles and fill in the table of values for the Hypotenuse sides of each of the triangles. You will find they are always a whole number value.

Opposite					
Adjacent	1	2	3	4	5
Hypotenuse					

Now use Pythagoras' Theorem to calculate the length of each of the Opposite sides. Write your answers to three decimal places.

Using the values in your table, now construct a second table by calculating the ratios of pairs of sides of the triangles.

$\frac{\text{Opposite}}{\text{Hypotenuse}}$					
$\frac{\text{Adjacent}}{\text{Hypotenuse}}$					
$\frac{\text{Opposite}}{\text{Hypotenuse}}$					

What do you notice about the ratios formed by these three sets of triangles? Write down your observations by completing the following statement:

For right angled triangles that are similar, that is ones that have the same set of angles but are different _____ , the ratios of the sides are _____ the _____.

Test this statement for triangles with angles other than 30°, 45° or 60°. You will need to be particularly careful with your measurements since most triangles will not give easy numbers to work with.

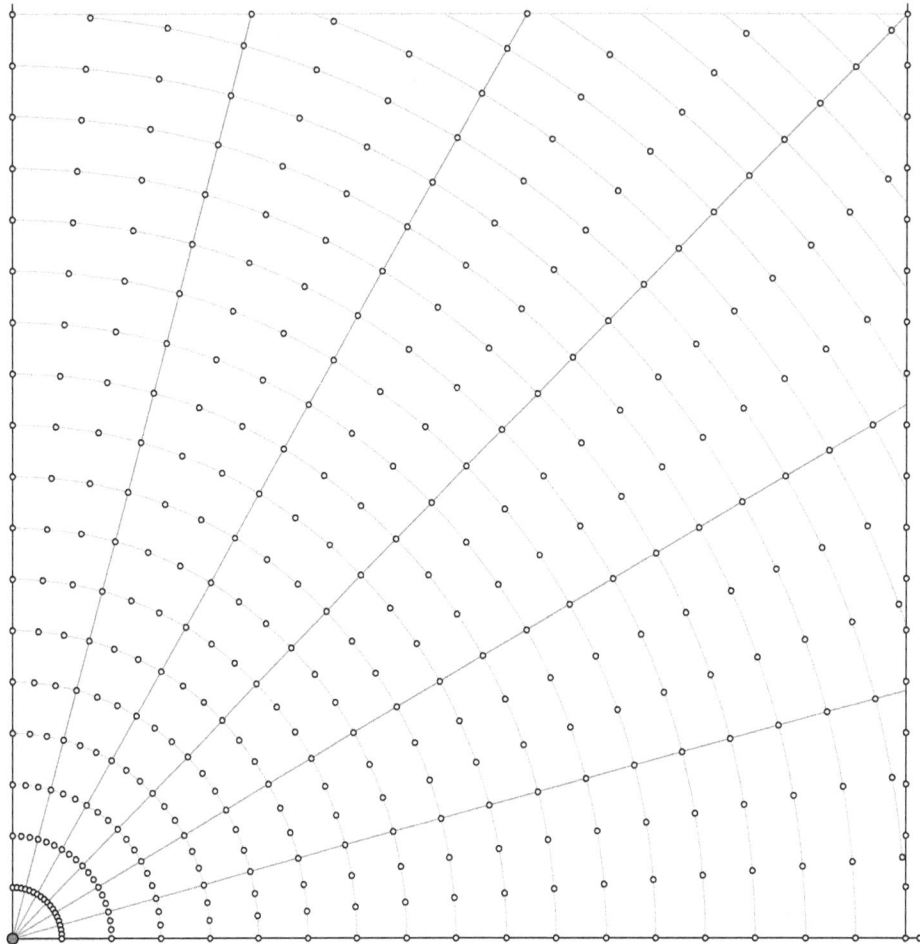

Discussion

Summarise your findings here:

For the sine ratio:

Any right triangle with the same angle has _____

As the size of the angle increases from 0° to 90° the sine ratio goes from _____ to _____

For the cosine ratio:

Any right triangle with the same angle has _____

As the size of the angle increases from 0° to 90° the cosine ratio goes from _____ to _____

For the tangent ratio:

Any right triangle with the same angle has _____

As the size of the angle increases from 0° to 90° the tan ratio goes from _____ to _____

Where to from here?

Now that you have a sense of what the sine, cosine and tangent ratios are, use your calculator to explore how much of a change in angle there needs to be to produce a change in the ratios by .1, .01, .001 and .0001. Start by looking at angles between 30° and 31°.

PYTHAGORAS AND TRIGONOMETRY
LESSON

4

Exploring Patterns in Triangle Ratios

A.C. Level:	9
A.C. Ref No's:	Apply trigonometry to solve right-angled triangle problems. ACMMG224
A.C. Substrands	Pythagoras and Trigonometry

Outcome

At the end of the activity students will know,

- The meanings of the terms 'adjacent' and 'opposite' sides in a right-angled triangle.

Students will be able to

- Calculate values for the three ratios, sine, cosine and tangent

- Select and accurately use the correct trigonometric ratio to find unknown sides (adjacent, opposite and hypotenuse) and angles in right-angled triangles.

Materials Required

CircleTrig paper

Right triangles are crucial in many parts of math, physics, and engineering. The parts of a right triangle are (not counting the right angle):

• two legs;

• one hypotenuse;

• two acute angles.

Finding the values of the side lengths and angles of a triangle by calculation rather than measurement is called solving a triangle.

You will soon work out how much information is needed to solve any right angled triangle. There is a certain minimum amount of information needed in order to complete this sort of task.

For each problem below, answer the question and work the example. Use each figure to keep track of what you know. Remember, you are using calculations from Pythagoras' Theorem and Trigonometry, not measurement to determine your answers. For this reason, the triangles are not drawn to scale.

1. Given one acute angle, what other parts can you find?
(Example: One acute angle is 21°.)

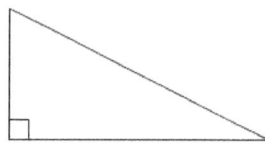

2. Given one side, what other parts can you find? (Example: One side is 3.)

3. Given two legs, what other parts can you find?
(Example: One leg is 4 and the other is 5.)

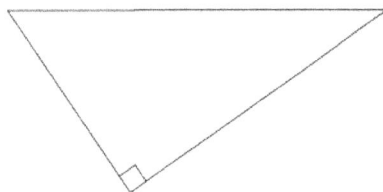

4. Given one leg and the hypotenuse, what other parts can you find?
(Example: One leg is 6 and the hypotenuse is 7.)

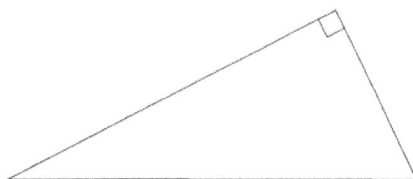

The figure at right shows one way to find an angle given an $\frac{opp}{hyp}$ ratio of 0.4.

By using the scales on the axes and the radius of the circle explain why the $\frac{opp}{hyp}$ ratio is 0.4 in this example.

Use a protractor to determine the angle formed at the left corner of the triangle.

You can use a similar method to find angles for given $\frac{adj}{hyp}$ ratios.

Enter your results in the table below.

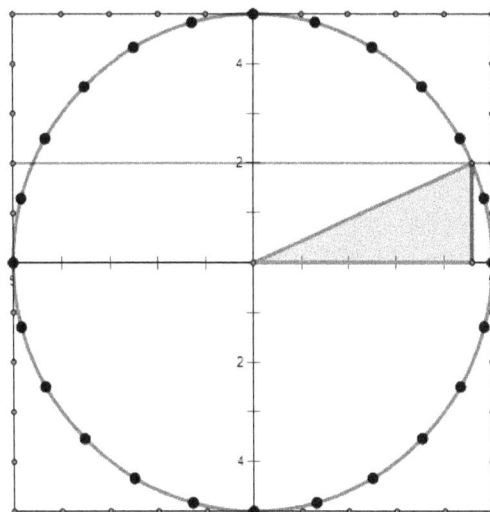

$\frac{opp}{hyp}$	θ
0	
0.2	
0.4	
0.6	
0.8	
1	

$\frac{adj}{hyp}$	θ
0	
0.2	
0.4	
0.6	
0.8	
1	

Discussion

Summarise your findings here:

A. When filling out the tables, look for patterns. What is the relationship between the ratios for complementary angles? For what angles do we have ratios of:

i)   0?

ii) 1?

B. Some of the triangles you used to fill out the tables are "famous right triangles". Check that the angles and ratios you found are correct by comparing your answers with those you found in Lesson 3.

C. Can the ratio $\frac{\text{opp}}{\text{hyp}}$ or the ratio $\frac{\text{adj}}{\text{hyp}}$ be greater than 1? Explain your answer.

Where to from here?

Pythagoras Theorem and Trigonometry can be used to solve problems whenever a right angled triangle can be seen or drawn. Describe at least five different types of problems where you might have to use these skills in real life. You can look at word problems in text books for some hints but put the ideas into your own words and explain why these techniques are appropriate.

PYTHAGORAS AND TRIGONOMETRY
LESSON

5

A New Area Formula

A.C. Level:	10 A
A.C. Ref No's:	Establish a formula for area using the sine ratio. ACMMG273
A.C. Substrands	Pythagoras and Trigonometry

Outcome

At the end of the activity students will know,

- The formula for area of a triangle involving the sine ratio

Students will be able to

- Calculate calculate area using this formula

- Show how this formula is connected to the one they knew previously involving base and height.

Materials Required

Mathomat

Activity 1

The use of trigonometry can be extended beyond right angled triangles in a number of ways. In this Lesson you will investigate how to make these extensions.

Area:

For many years you have known that the formula for the area of a triangle is:

$$\text{Area} = \frac{1}{2} \times \text{base} \times \text{height}$$

This formula requires that we be given both the base length and the perpendicular height of the triangle.

What happens, though, when the height is not given but a couple of the sides and one of the angles?

Start by drawing a line 10 cm long starting from the point A. Mark the other end as B.

A

Next, from point A, measure a second distance 6 cm. Mark this point as D.

Now use your protractor to mark an angle of 60°, again from point A. Draw a line at this angle for a length of 6 cm. Mark this end as B.

Complete the triangle by joining points B and C together.

We now have a triangle where we have been given measurements of the base and one other side of the triangle and one of the angles, but not the height. While it is possible in this instance to measure the height of the triangle, it is not always so in real problems.

Using just the angle and the side AB it is possible to calculate the value of the height of the triangle. Complete the formula for finding height.

Height =

Now replace height in the area formula: $Area = \frac{1}{2} \times base \times$ _____

Use this formula to work out the area of the triangle.

Discussion

What is the link between the two formulas for the area of a triangle?

In what circumstances would you use each formula?

Where to from here?

There will be times when you are given the area of a triangle and need to work out side lengths, height or even an angle of the triangle. Using the two area formulas you now know find new formulas for solving each of these problems. Show all the steps in your working so that others can understand your thought process.

Check and share your formulas with others in the class.

PYTHAGORAS AND TRIGONOMETRY LESSON

6

The Sine Rule

A.C. Level:	10 A
A.C. Ref No's:	Establish the sine rule for any triangle and solve related problems ACMMG273
A.C. Substrands	Pythagoras and Trigonometry

Outcome

At the end of the activity students will know,

- The formula for the Sine Rule

Students will be able to

- Calculate calculate lengths and angles using this formula

- Show how this formula is developed from right angled trigonometry

Materials Required

Mathomat

Activity

Sine Rule

Write down the expression for height that you worked out for the area formula.

Height =

Now look at the triangle on the right.
None of the angles are right angles.

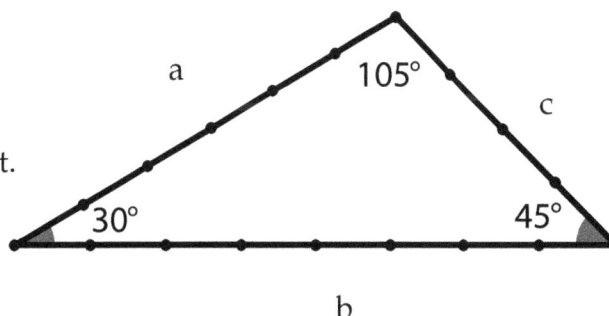

It is possible, though, to break it into two smaller triangles, by adding in the vertical height. This gives a diagram very similar to the one used to find the area of a triangle in the last activity.

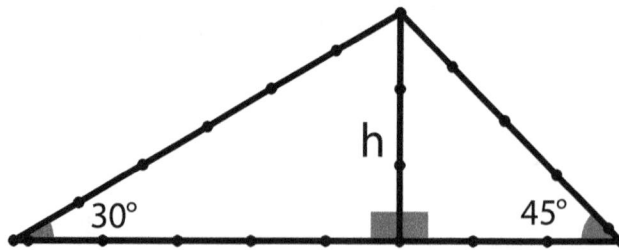

Using the dots to count the lengths rather than measure them, You will find that for the triangles drawn here two of the sides in each triangle are whole number values but the third one is not. Use Pythagoras to find the exact value of the third side and write as a surd,

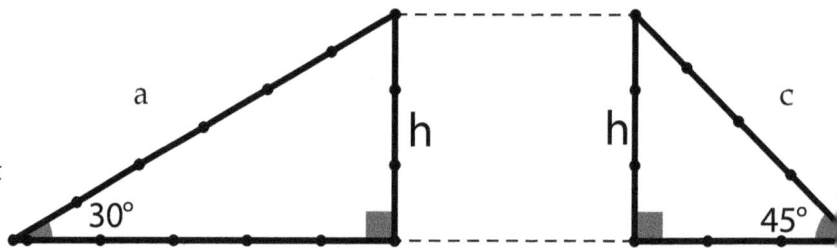

—

For the triangle on the left side of the diagram, the one with the 30° angle, use your height formula to write down an expression for h.

h = _____

Now use the triangle on the right of the diagram, with 45° to write a second expression for h.

h = _____

These two expressions are referring to the same value, h. This means that we can put them together and not have to use h at all.

Left hand expression   Right hand expression

_____= h = _____

Now remove h from the middle bit:

Left hand expression   Right hand expression

_____ = _____

Now, instead of using the lengths of the two hypotenuses, ie 6 and 3 $\overline{2}$, use the letter labels a and c.

Left hand expression   Right hand expression

_____ = _____

Finally, instead of using the angle values of 30° and 45° use the labels C and A.

Left hand expression   Right hand expression

_____ = _____

Questions: Look carefully at the last two instructions.

1) What is different about the labels you were told to use?

2) What is different about the order of the labels you were told to use?

Discussion

Discussions and key points:

The Sine Rule works for any triangle, not just right angled ones. What do we do to the triangle in order to make this possible?

Which measurement or calculation is needed to establish the rule but then disappears from the final version of it? Explain why this happens.

Where to from here?

There are some occasions where the Sine Rule gives two answers. These answers mean that there are two possible triangles described by the original information given. In each case the angle given is an acute angle.

Explore these "ambiguous cases". Start by drawing two triangles where the angle A° = 30°, side a = 5 cm and side b = 8 cm.

PYTHAGORAS AND TRIGONOMETRY LESSON

7

The Cosine Rule

A.C. Level:	10 A
A.C. Ref No's:	Establish the cosine rule for any triangle and solve related problems ACMMG273
A.C. Substrands	Pythagoras and Trigonometry

Outcome

At the end of the activity students will know,

- The formula for the Sine Rule

Students will be able to

- Calculate calculate lengths and angles using this formula

- Show how this formula is developed from right angled trigonometry

Materials Required

Mathomat

Extending Pythagoras Theorem to Non-Right Angled Triangles

You have worked with Pythagoras Theorem before and know that for right angled triangles there is a direct relationship between the lengths of the three sides. It is summarised in the formula:

$$c^2 = a^2 + b^2$$

Where a and b are the shorter two sides of the triangle and c is the longest side.

In Lesson 1 of this workbook you found that the formula only works for right angled triangles.

For Acute angled triangles: $c^2 < a^2 + b^2$

While for Obtuse angled triangles: $c^2 > a^2 + b^2$

There is enough of a pattern here to ask the question of whether there are equations that will work for triangles other than right angled ones.

This activity is significantly different to any of the others in this series. It asks you, the student, to focus your attention strongly on a set of specific examples looking for and then testing patterns to discover new connections between the parts of a triangle. At the end of the process you will have a single formula that works for all triangles, not just right angled ones.

The purpose of this activity is not just to give you a new formula, but to lead you through the thinking process that goes on inside someone's head when they are discovering something new.

Step 1: Start with a right angled triangle

Use the now familiar 3, 4, 5 triangle as shown in the diagram.

Let a = 3, b = 4 and c = 5.

Use the Pythagorean Formula to check that $a^2 + b^2 = c^2$.

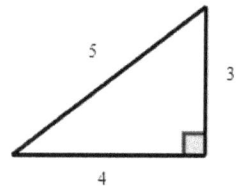

Step 2: Now use the triangle with a = 3, b = 4 and c = $\overline{13}$.

Where the right angle used to be there is now a 60° angle.

$a^2 + b^2$ has not changed, but c^2 has.

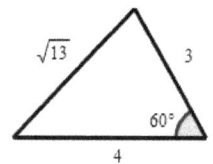

Write down the value of c^2

Find the difference between $a^2 + b^2$ and c^2.

$a^2 + b^2 - c^2 =$ _____

Using just the values of 3 and 4, can you write down an expression that gives c^2.

_____

Now replace 3 and 4 with a and b to give a generalised expression (ie something that looks like a formula.)

_____

Step 3: Test the formula you have just made.

For a third triangle, with c = 2.83363 and an angle of 45° try
the approach you just used in step 2.

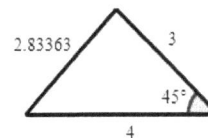

It should be little surprise that this time the approach did
not work. That is alright. It just means there is more to the
relationship than what we have so far.

Try this hint: sin(45) and cos(45) both equal $\frac{1}{2}$. . Try adding this value to your
calculation to show that it makes a difference.

_____

Now write down the two possible "formulas" that worked for this triangle.
Remember to use a, b and c for the lengths and θ for the angle.

$c^2 =$ _____

Step 4: Test both formulas to see which one is correct.

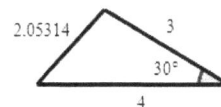

This time use the triangle with c = 2.05314 and angle θ = 30°.

Find the values of:

sin 30° = _____

cos 30° = _____

Use these values in your "formulas" from Step 4:

sin formula: _____

cos formula: _____

Which formula works? _____

Now go back to the triangle in Step 2 and use your formula to check.
Show your working here.

Discussion

Discussions and key points:

The Cosine Rule is Pythagoras' Theorem with an added term for non-right triangles. Complete the expression to give this third term.

$$- a \times b \times \underline{\hspace{2cm}}$$

Now write the full version of the Cosine Rule:

$$c^2 = a^2 + b^2 - a \times b \times \underline{\hspace{2cm}}$$

Where to from here?

The Cosine Rule can also be used to find the value of an unknown angle, given the values of the three sides. Rearrange the formula above to show this new formula.

$$C° = \underline{\hspace{3cm}}$$

PYTHAGORAS AND TRIGONOMETRY
LESSON

8

The Unit Circle

A.C. Level:	10 A
A.C. Ref No's:	Use the unit circle to define trigonometric functions, and graph them with and without the use of digital technologies ACMMG274
A.C. Substrands	Pythagoras and Trigonometry

Outcome

At the end of the activity students will know,

* How the unit circle allows for calculation of trigonometric values of angles greater than 90°

Students will be able to

* Calculate the sine and cosine values for angles between 0° and 360°.

* Graph sine and cosine functions

Materials Required

Calculator

The Unit Circle

This activity focuses on looking for patterns in the numbers produced by the sine and cosine ratios.

Examine the Radial Grid below. Angles are marked every 10° from 0° to 90° and also at 45°. A triangle with a left angle of 30° has been drawn so that the hypotenuse is 10 cm in length.

Check by measuring, that the lengths of the opposite and adjacent sides are 5 cm and about 8.6-8.7 cm respectively.

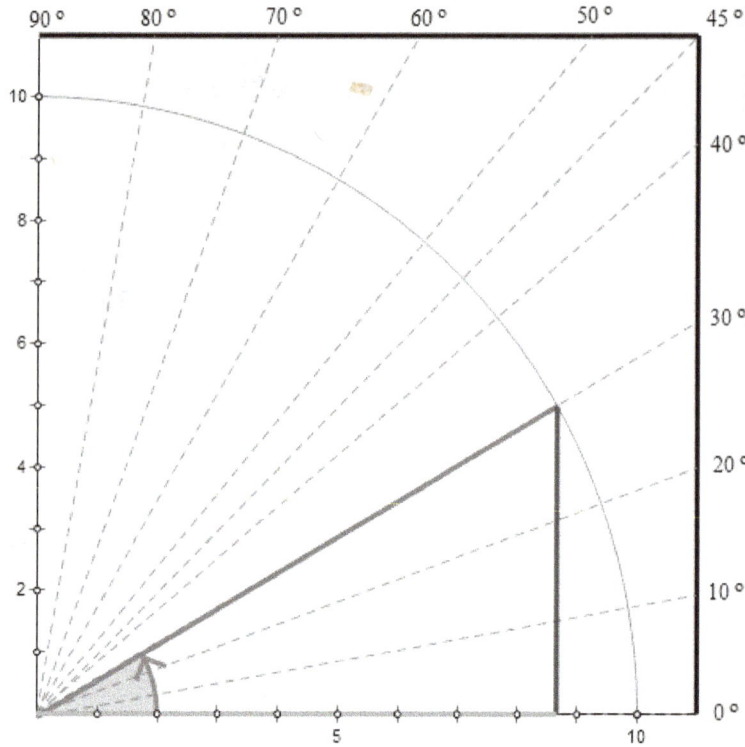

This in turn means that the sine and cosine ratios for 30° are 0.5 and 0.86. Explain why this is so.

Now, using measurement or calculation, fill in the table. You can draw the triangles onto the grid if it helps.

Questions:

1) What do you notice about the two lists of numbers for sine and cosine?

2) Can you find pairs of angles, call them a and b, where:

   sin a + cos b = 0?

3) What is the connection between the angles a and b in these cases?

Angle	sin	cos
0°		
10°		
20°		
30°	0.5	0.866
40°		
45°		
50°		
60°		
70°		
80°		
90°		

In order to tell the difference between this triangle and the one on the precious page, we can use the axes of a Cartesian graph and mark the adjacent side as -8.6.

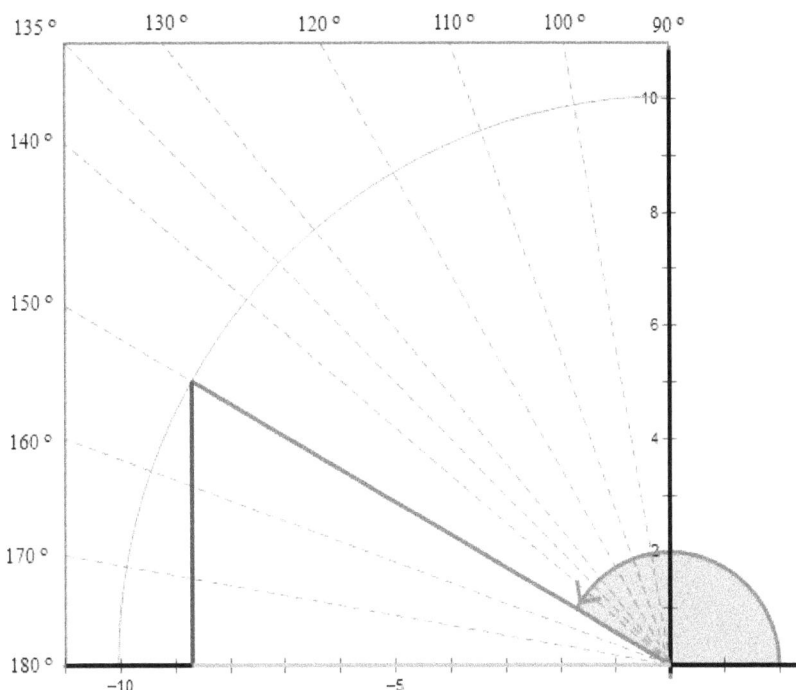

Complete the table for this second quadrant.

Questions:

If we continue this pattern what will happen for angles between 180° and 270° for:

a) the sine ratio:
    (Hint, are the values on the axis positive or negative?)

b) the cosine ratio:
    (Hint, are the values on the axis positive or negative?)

If we continue this pattern what will happen for angles between 270° and 360° for:

a) the sine ratio:
    (Hint, are the values on the axis positive or negative?)

b) the cosine ratio:
    (Hint, are the values on the axis positive or negative?)

Angle	sin	cos
90°		
100°		
110°		
120°		
130°		
135°		
140°		
150°	0.5	-0.866
160°		
170°		
180°		

Now, fill in the remaining values:

Angle	sin	cos
90°		
100°		
110°		
120°	-0.5	-0.866
130°		
135°		
140°		
150°		
160°		
170°		
180°		
220°		
225°		
230°		
240°		
250°		
260°		
270°		
280°		
290°		
300°		
315°		
320°		
330°	-0.5	-0.866
340°		
350°		
360°		

Once you have a full set of values, use them to complete the sine graph on the next page. The points for 30°, 45° and 60° have been done for you.

Once you have completed the sine curve, you should be able to produce a cosine curve as well using the second set of axes that run down the page. Remember that the cosine value of an angle is the length of the adjacent side of the triangle.

Explain why the axes for the cosine curve are set at 90° to the axes for the sine curve.

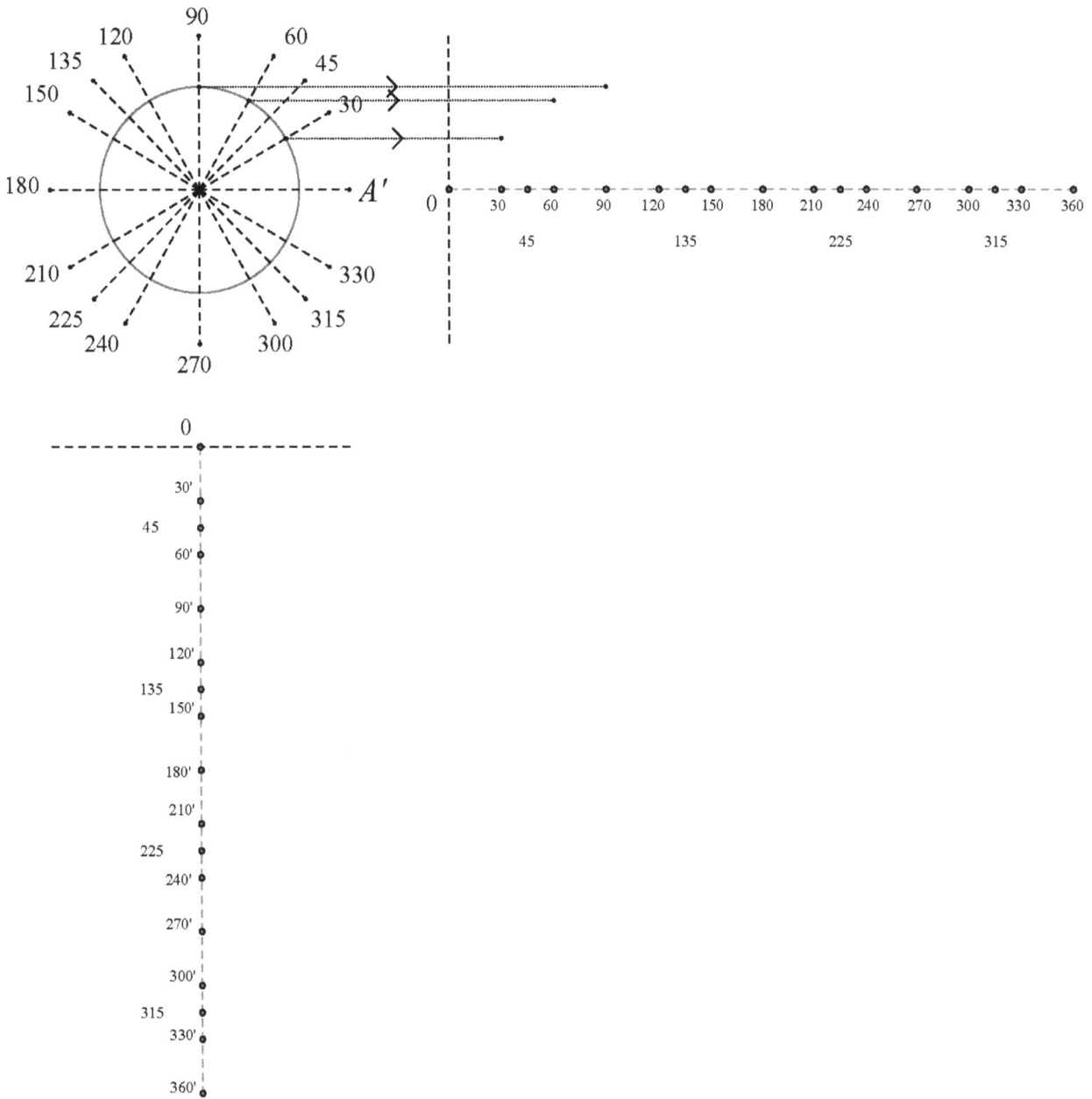

Discussion

The unit circle makes it possible to find ratios of sine, cosine and tangent for any angle between 0° and 360° while only ever using right angled triangles. Explain how this is possible.

In graphing the sine and cosine curves, the angle size goes along the independent axis and the ratio along the dependent axis. We often call these curves "waves". Why?

Where to from here?

What would the graphs of 2 *sin(x)* or 3 *cos(x)* look like? Try drawing these curves by hand first then use a calculator or computer to check your results.

What does changing the number in front of the ratio do to the size and shape of the curve?

GEOMETRIC REASONING LESSON

1

Observing angles around a point

A.C. Level:	5
A.C. Ref No's:	Estimates, measures and compares angles using degrees. Constructs angles using both a 180° and 360° protractor. ACMMG112
A.C. Substrands	Geometric Reasoning

Outcome

At the end of the activity students will know,

- That angles are a measure of turn

- That only some types of quadrilaterals will neatly fill around a point

Students will be able to

- Use protractors to measure angle sizes in degrees

Materials Required

Mathomat

1. First find the following shapes on your Mathomat

 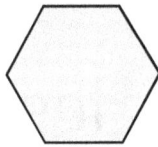 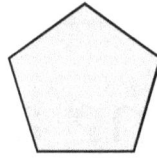 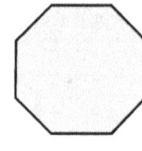

| Square | Hexagon | Pentagon | Rhombus | Octagon | Trapezium |

2. Then do the worksheet on the next page

3. Summarise your findings in the table below:

Table 1.1

Shape	Number of times	Shape	Number of times
Square		Rhombus	
Hexagon		Octagon	
Pentagon		Trapezium	

Draw and observe angles around a point

Using your Mathomat find out how many times the six polygons illustrated on the previous page will fit around a point. Summarise your findings in Table 1.1 (on the previous page).

One of the shapes has already been done for you.

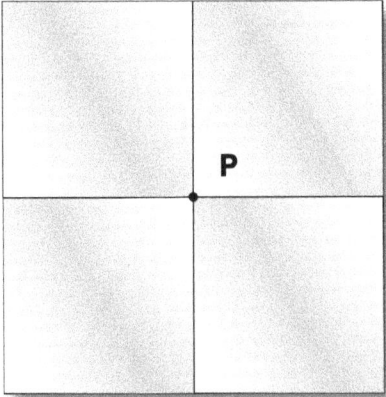

P

• P

• P

• P

• P

• P

Constructing a Protractor

Materials Required

The Mathomat template and a piece of string

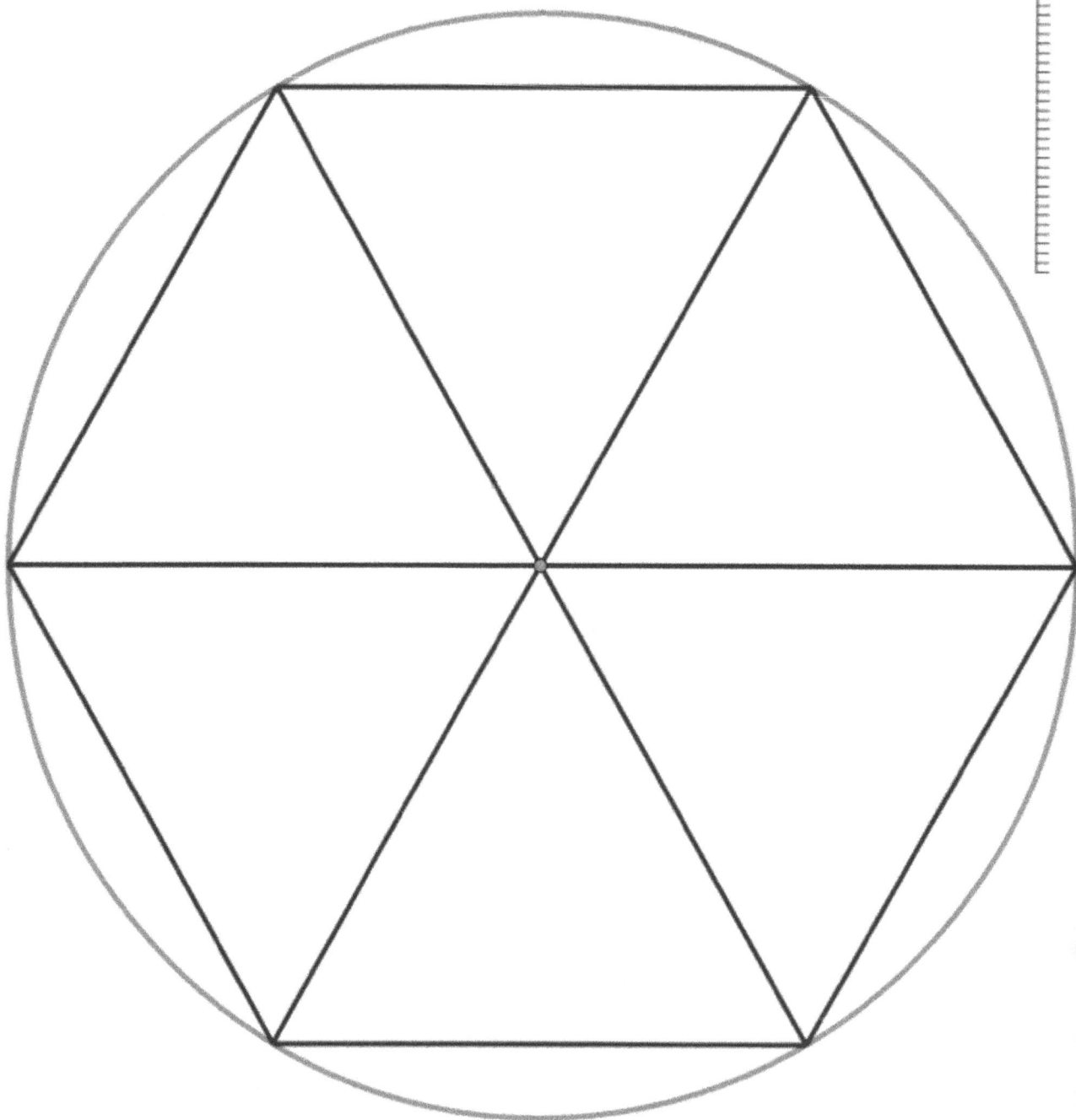

Step 1: Construct a regular hexagon using a Mathomat and the circle provided.

Step 2: Line up a piece of string with the 90 mm line on the right side of the page. Mark the string with 1.5 mm divisions.

Step 3: Place the string carefully around the circumference of the circle. Use the markings to produce 1° gradations around the circumference.

Notes on the origin of the use of 360° in a circle can be found in the appendix.

Find the Polygons listed in the shaded box below on your Mathomat, draw them in the space below and use your Mathomat protractors to measure their external angles. Put your findings in Table.

(One of the shapes has already been done for you)

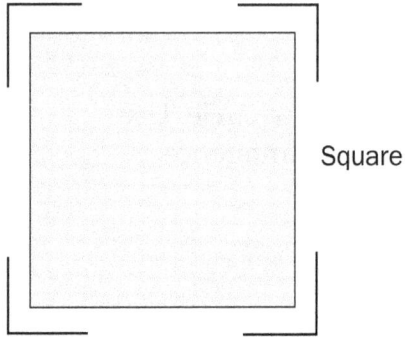

Square

• Square
• Hexagon
• Octagon
• Pentagon
• Trapezium
• Rhombus
• Triangle

Consider this question while working:
How would you order these shapes based on angle size?

Summarise your findings in the table below:

Shape	Number of times	Shape	Number of times
Square		Rhombus	
Hexagon		Octagon	
Pentagon		Trapezium	

Discussion

Questions and key points

- What is our unit of measurement for angle?

- Where did it come from?

- Why 360° rather than 100°?

Where to from here?

- This investigation opens up exploration, analysis and reasoning opportunities such as: why do some angles fit evenly around a point while others don't?

- Can we use this property to define or set an agreed unit or measurement for angles?

(This question leads in to the next investigation.)

GEOMETRIC REASONING LESSON

2

Angle Measurement

A.C. Level:	5
A.C. Ref No's:	Estimates, measures and compares angles using degrees. Constructs angles using both a 180° and 360° protractor. ACMMG112
A.C. Substrands	Geometric Reasoning

Outcome

At the end of the activity students will know,

- That angles are a measure of turn

- That only some types of quadrilaterals will neatly fill around a point

Students will be able to

- Use protractors to measure angle sizes in degrees

Materials Required

Mathomat

1. What are the measures of the angles that share a vertex at the centre of

a. A Chrysler symbol?

b. A Mercedes symbol?

c. A peace sign?

d. A clock, between consecutive hours?

e. A cross?

2. Find the measures of all the angles for each of the polygons shown below.
    Write the angle measures in the shapes.

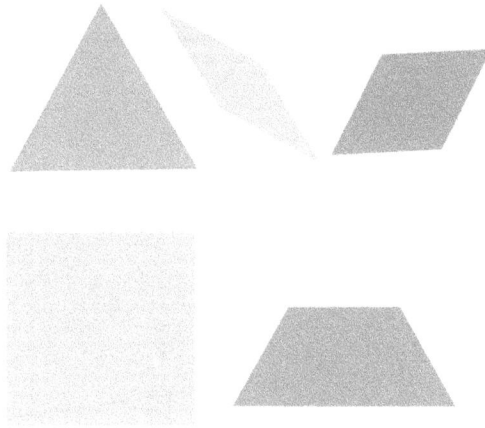

3. One way to measure angles is to place smaller angles inside larger ones. For example, six copies of the small angle on the thin rhombus fit inside the figure below.

This figure, called a protractor, can be used to measure all pattern block angles.

a. Mark the rest of the lines in the figure with numbers.

b. Use it to check the each of the angles on the given polygons.

c. Using the polygons, add the 15° lines between the 30° lines shown in the protractor.

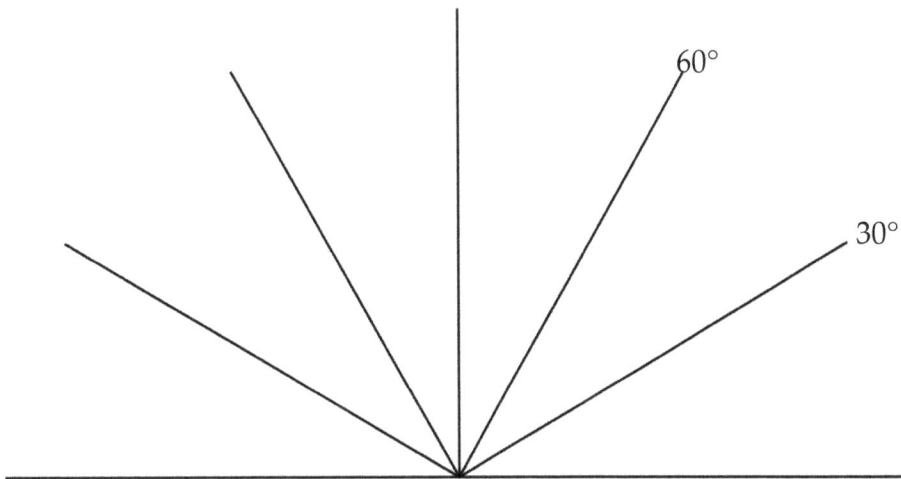

60°

30°

4. Use the protractor on your Mathomat to measure the angles on the triangles below. For each one, add up the angles. Write the angle sum inside each triangle.

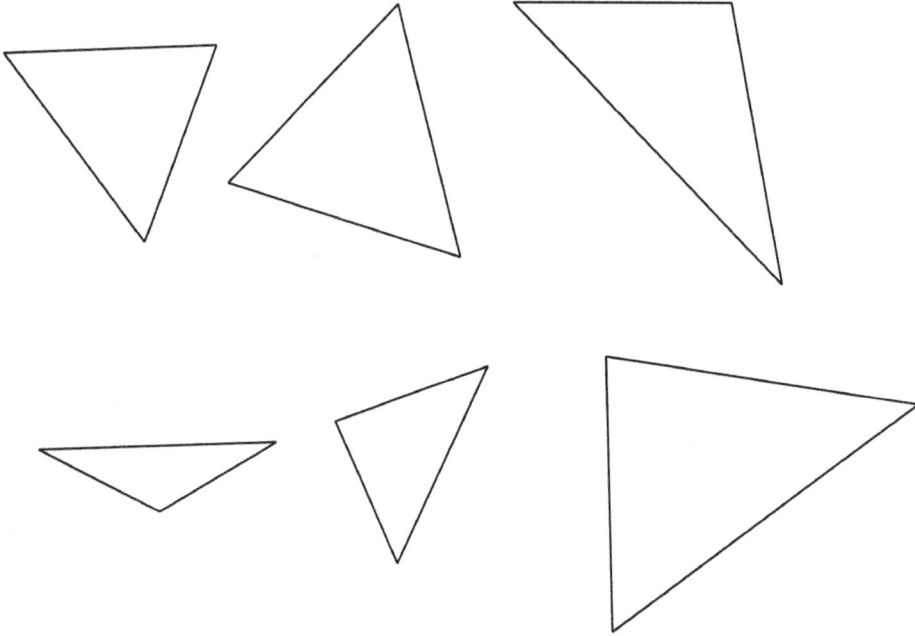

5. Use the protractor on your Mathomat to draw the following angles in the space below.

a. 20°

b. 50°

c. 100°

Discussion

Questions and key points

Which of the parts in question 1 did you find most difficult to answer?

What made the others easier?

Where to from here?

The world we live in is made up of all types of different shapes. In designing and building everything the objects we use everyday, artists, designers, builders and engineers have to measure and construct angles all the time.

Name, draw and measure the angles of other shapes you can see around you.

GEOMETRIC REASONING LESSON

3

Clock Angles

A.C. Level:	6
A.C. Ref No's:	Investigate, with and without digital technologies, angles on a straight line, angles at a point and vertically opposite angles. Use results to find unknown angles. Define acute, obtuse, reflex and right angles. Identify that angles have arms and a vertex and that the size is the amount of turn required for one arm to coincide with the other. The angle is measured in degrees with a protractor ACMMG141
A.C. Substrands	Geometric Reasoning

Outcome

At the end of the activity students will know,

- That angles are a measure of turn

- That only some types of quadrilaterals will neatly fill around a point

Students will be able to

- Use protractors to measure angle sizes in degrees

Materials Required

Mathomat

This activity requires you to put to use what you know about both angles and clocks.

Consider the following questions:

1   How many degrees does the hour hand travel in an hour? In a minute?

2   How many degrees does the minute hand travel in an house?
     In a minute?

3   What angles doe the hour and minute hands of a clock make with each other at different times? Write an illustrated report.

Start by figuring it out on the hour (for example at 5:00).
Then see if you can answer the question for the half-hour
(for example, at 5:30).
Continue exploring increasingly difficult cases.

Remember that the hour hand moves, so, for example, the angle at 3:30 is not 90° but a little less, since the hour hand has moved halfway towards the 4.

Use the clocks on the next page for practice.
Cut and paste clocks into your report.

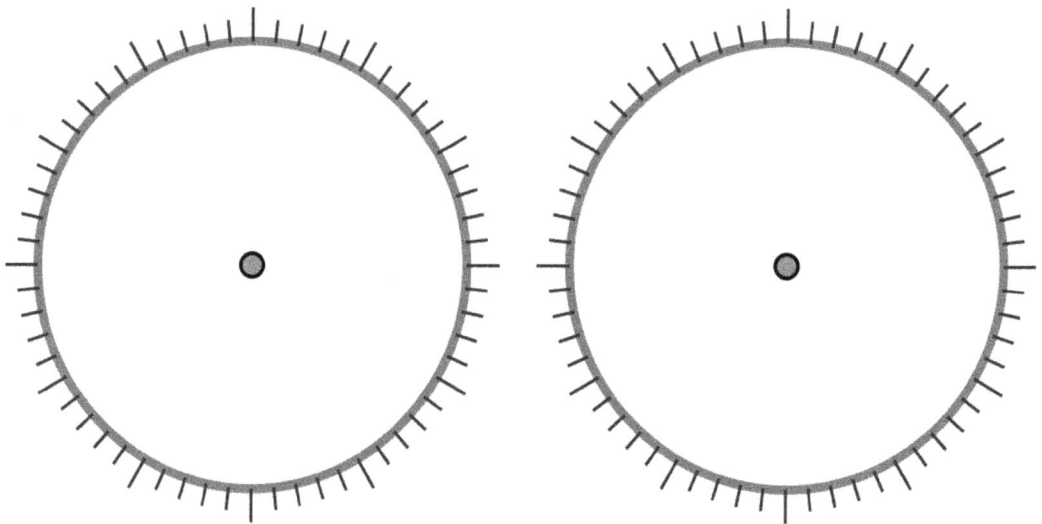

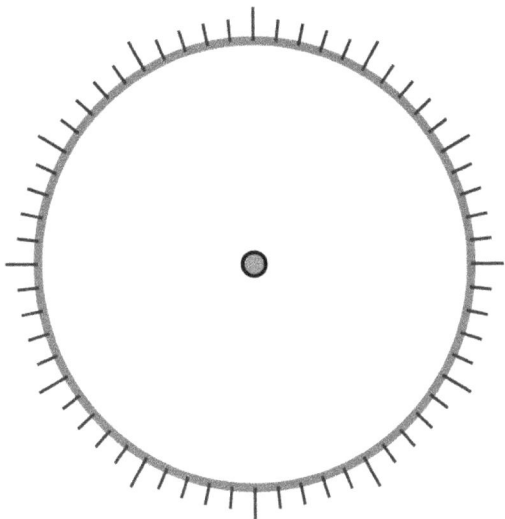
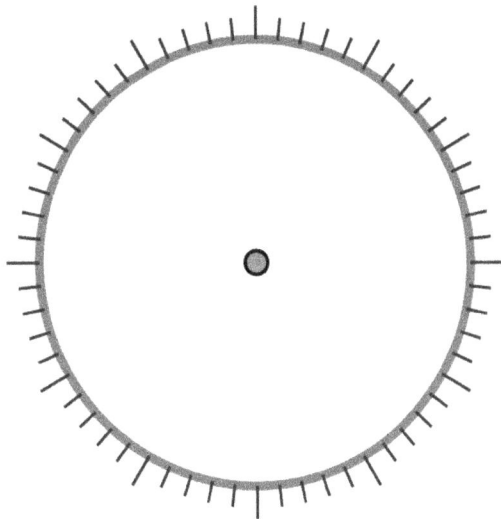
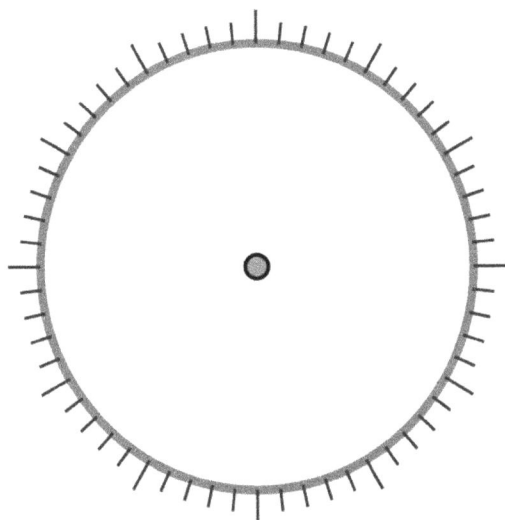
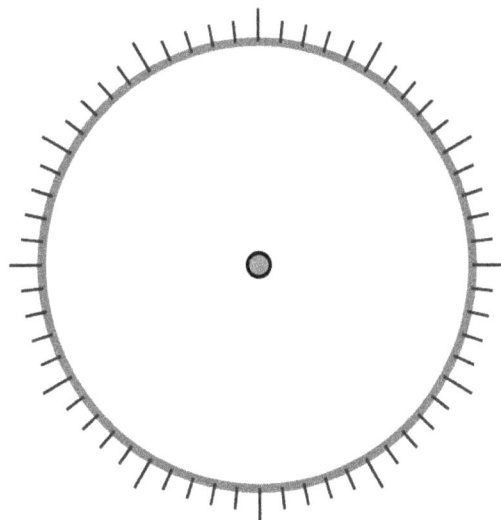
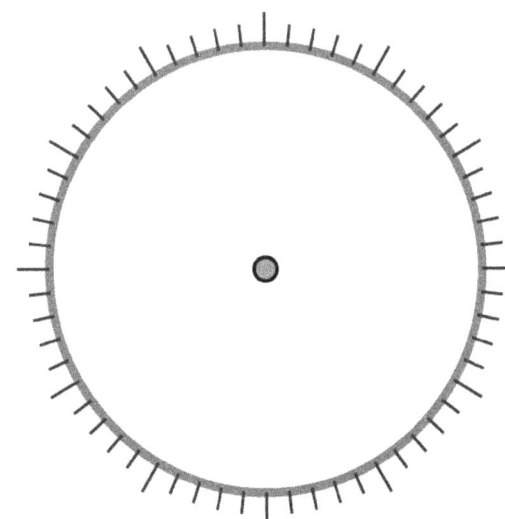
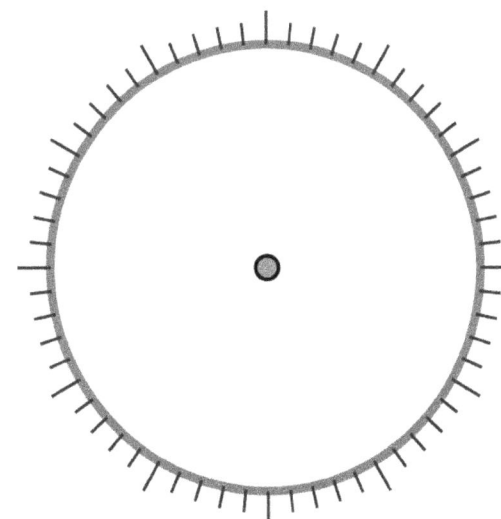

Discussion

Questions and key points

Classify each of your angles as acute, obtuse, reflex, right or straight angles.

Of course, the hands of a clock always produce two angles. Discuss in groups and then with the class, how you decided on which angle to measure for each time.

Where to from here?

Go back to your clock faces and add in the values and the names of the second angle in each diagram.

What pattern develops between pairs of angles you have now marked?
For example, an acute angle always pairs up with what other type of angle?

GEOMETRIC REASONING LESSON

4

Angles of Mathomat Polygons

A.C. Level:	7
A.C. Ref No's:	Demonstrate that the angle sum of a triangle is 180° and use this to find the angle sum of a quadrilateral ACMMG163
A.C. Substrands	Geometric Reasoning

Outcome

At the end of the activity students will know,

• That angles are a measure of turn

• That only some types of quadrilaterals will neatly fill around a point

Students will be able to

• Use protractors to measure angle sizes in degrees

Materials Required

Mathomat

Equipment

Mathomat Shapes

1. Find the sum of the angles for each shape, and record them below

Triangle (#23)	
Square (#1)	
Rhombus (#16)	
Rhombus (#19)	
Trapezoid (#24)	
Hexagon (#9)	

2. Which shapes have the same sum?

3. Use templates to make a polygon such that such that the sum of its angles is the same as the sum for the hexagon.

Sketch your solution in the space at right. How many sides does it have?

4. Use the templates to make a polygon such that the sum of its angles is less than the sum for the hexagon but more than the sum for the square.

Sketch your solution in the space on the next page.
How many sides does it have?

5. Use the templates to make a polygon such that the sum of its angles is greater than the sum for the hexagon. Sketch your solution in the space at right.

How many sides does it have?

6. For each number of sides from 3 to 12, make a polygon and find the sum of its angles. Sketch your solutions on a separate sheet of paper and fill out the table below.

Sides	Sum of the angles
3	
4	
5	
6	
7	

Sides	Sum of the angles
8	
9	
10	
11	
12	

Discussion

Questions and key points

If a polygon had 20 sides, what would be the sum of its angles? Explain.

If the sum of the angles of a polygon were 4140°, how many sides would it have? Explain.

If the sum of the angles of a polygon were 450°, how many sides would it have? Explain.

Where to from here?

What is the relationship between the number of sides of a polygon and the sum of its angles? Write a sentence or two to describe the relationship, or give a formula.

GEOMETRIC REASONING LESSON

5

Classifying Triangles

A.C. Level:	7
A.C. Ref No's:	Classify triangles according to their side and angle properties of scalene, isosceles, right-angled and obtuse-angled triangles ACMMG165
A.C. Substrands	Geometric Reasoning

Outcome

At the end of the activity students will know,

*   The names and definitions for the different types of triangles.

Students will be able to

*   Draw examples of each type of triangle and identify right angled triangles by their angle values.

Materials Required

Mathomat

1. Could you have a triangle with two right angles? With two obtuse angles? Explain.

2. For each type of triangle listed below, give two possible sets of three angles (in some cases, there is only possibility).

   a. Equilateral:

   b. Acute isosceles:

   c. Right isosceles:

   d. Obtuse isosceles:

   e. Acute scalene:

   f. Right scalene:

   g. Obtuse scalene:

3. If you cut an equilateral triangle exactly in half, into two triangles, what are the angles of the "half-equilateral" triangles?

4. Which triangle could be called "half-square"? What are the angles in this triangle?

5. Explain why the following triangles are impossible:

   a.   Right equilateral

   b.   Obtuse equilateral

6. Among the triangles listed in question 2, which have a pair of angles that add up to 90°?

7. Make up two more examples of triangles in which two of the angles add up to 90°. For each example, give the measures of all three angles.

8. Complete the sentence:

   "In a right triangle, the two acute angles _____"

9. Trace all the triangles on the template in the space below and label them by type:

equilateral (EQ)	acute isosceles (AI)
right isosceles (RI)	obtuse isosceles (0I)
acute scalene (AS)	right scalene (RS)
half-equilateral (HE)	obtuse scalene (OS)

Discussion and Wrap up

Questions and key points

What is unique about right angled triangles?

What is unique about equilateral triangles?

These two types of triangles are particularly useful. Give reasons why this might be so.

Where to from here?

Triangles are used in building all types of structures. This is because they hold their shape well under pressure. One example of such a structure is called a "geodesic dome". Find out how these domes are made and what type of triangle is used.

Would other types of triangles work as well or better? Explain why. You could demonstrate your conclusions by actually making domes with different types of triangles using straws or pipe cleaners.

GEOMETRIC REASONING LESSON

6

The Exterior Angle Theorem

A.C. Level:	7
A.C. Ref No's:	This concept is not specified in the Australian Curriculum. It is connected to the properties of triangles and is included here as an extension activity.
A.C. Substrands	Geometric Reasoning

Outcome

At the end of the activity students will know,

- The definition of an exterior angle of a triangle.

- The relationship between exterior and interior angles of triangles.

Students will be able to

- Calculate the exterior angles of any triangle

Materials Required

Mathomat

In this figure, angle BAC = 60° and angle ABC = 40°. One side of the angle at C has been extended, creating an exterior angle:

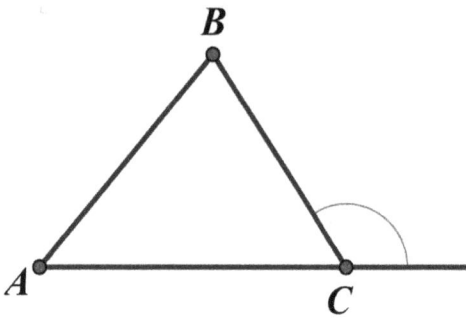

Without measuring, since this figure is not to scale, find the interior angle at C (ACB) and the exterior angle at C (BCD).

Interior angle at C _

Exterior angle at C _

Make up two other examples of triangles where two of the angles add up to 100°. Write all three angles for each triangle in the spaces below. Sketch the triangles in the space below and at right.

3.   For each of your triangles, find the measures of all three exterior angles.

Triangle one: _

Triangle two: _

4.   In the space at right, sketch an example of a triangle ABC where the exterior angle at B is 65°.

a. What is the measure of the interior angle at B?

b. What is the sum of the other two interior angles (BAC and ACB)?

5.   Repeat Problem 4 with another example of a triangle ABC with a 65° exterior angle at B.

a. What is the interior angle at B?

b. What is the sum of the other two interior angles (BAC and ACB)?

6.   I am thinking of a triangle ABC. Exterior angle A is 123°. Interior angle B is given below. What is the sum of interior angles B and C? (Use a sketch to help you work this out.)

a. If angle ABC = 10°

b. If angle ABC = 20°

c. Does it matter what angle ABC is? Explain.

7. I am thinking of a triangle ABC. Exterior angle A is O.
    What is the sum of interior angles B and C?
    Explain how you get your answer.

8. Complete this statement and explain why you think it is correct:
    The Exterior Angle Theorem: An exterior angle of a triangle is always
    equal to...

9. What is the sum of the two acute angles in a right triangle?
    Is this consistent with the exterior angle theorem? Explain.
    (Hint: What is the exterior angle at the right angle?)

10. What is the sum of all three exterior angles of a triangle?
    Find out in several examples, such as the ones in Problems 1, 2, 4 and 5.
    Explain why the answer is always the same.

Discussion

Questions and key points

How is the exterior angle of a triangle related to the internal angles?

How many exterior angles does every triangle have?
Draw a diagram to show where they are.

Where to from here?

Other polygons also have exterior angles. Is there a clear simple relationship between the exterior angles of other polygons and their interior angles? Explain, giving reasons why or why not.

GEOMETRIC REASONING LESSON

7

Angles and Triangles in a Circle

A.C. Level:	10A
A.C. Ref No's:	Prove and apply angle and chord properties of circles ACMMG272
A.C. Substrands	Geometric Reasoning

Outcome

At the end of the activity students will know,

- Which types of triangles can and cannot be made with one vertex at the centre of a circle and the other two on the circumference.

Students will be able to

- Draw and describe the properties of triangles inside a circle.

Materials Required

Circle Geoboard, Mathomat, Geometers Sketchpad

Types of triangles

This first section of the activity relates back to the investigation of triangle types in year seven. The additional idea is that all of the triangles are now drawn inside the confines of a circle. This will bring to light new properties based on the combination of the two shapes.

1 Make triangles on the circle geoboard, with one vertex at the centre and the other two on the circle. (See example below)

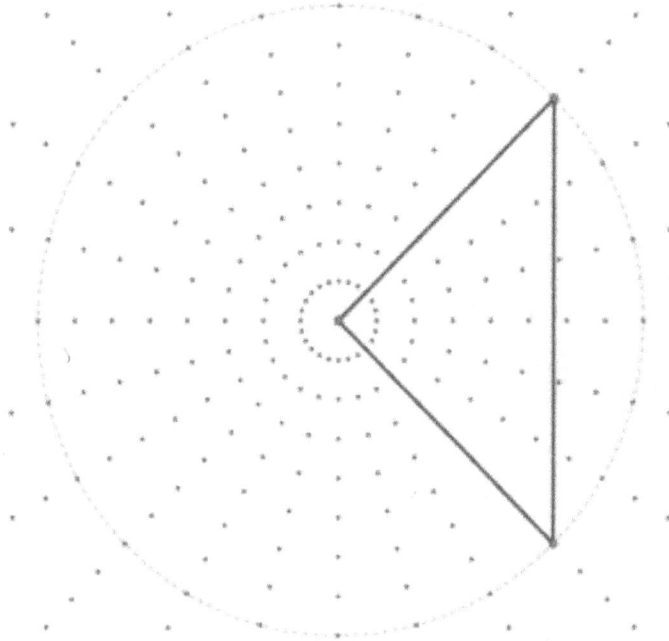

Equilateral (EQ)

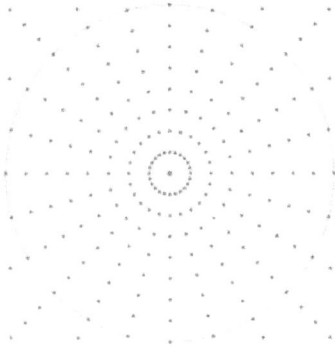

Right isosceles (RI)

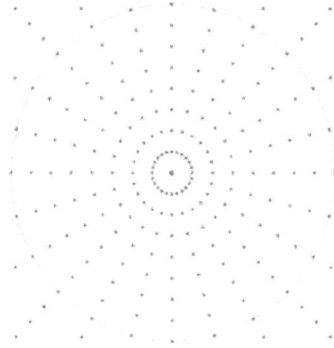

Acute scalene (AS)

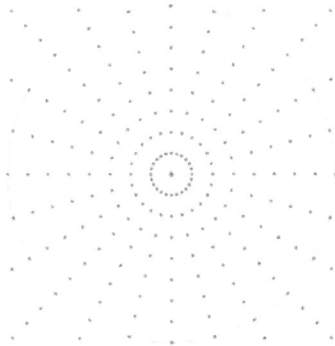

Half-equilateral (HE)

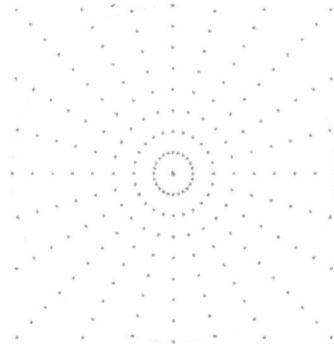

Acute isosceles (AI)

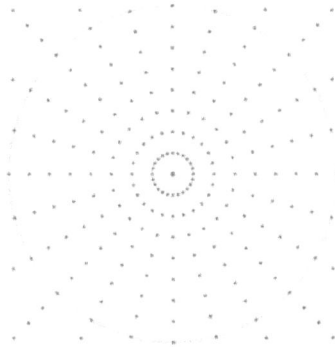

Obtuse isosceles (OI)

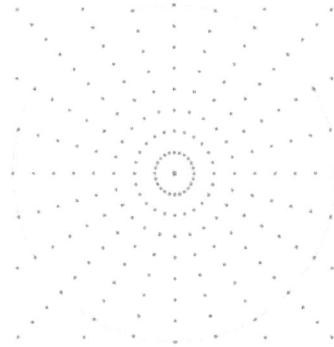

Right scalene (RS)

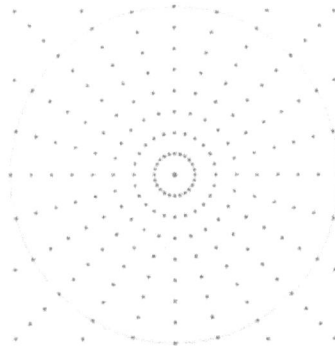

Obtuse scalene (OS)

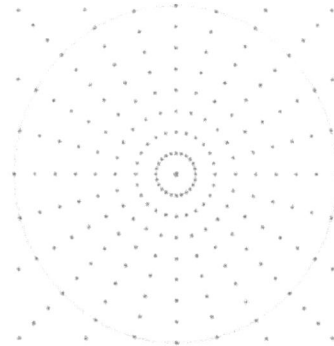

2 a. If possible, make one of each of the eight types of triangles listed above. You do not have to do them in order!

b. Sketch one of each of the types of triangles it was possible to make on circle (geoboard paper). Identify which type of triangle it is, and label all three of its angles with their measures in degrees.
(Do not use a protractor! Use geometry to figure out the angles.)

Thinking back

3 a. What is true of all the possible triangles in Problem 1

b.   Summarize your strategy for finding the angles. Describe any shortcuts or formulas you used.

Definition

A triangle is inscribed in a circle if all three of its vertices are on the circle.

Repeat Problem 1 with inscribed triangles such that the circle's center is on a side of the triangle. (See circle 2.) Hint: Drawing an additional radius should help you find the measures of the angles.

4. Thinking back:

a. Summarize your strategy for finding the angles in Problem 3. Describe any shortcuts or formulas.

b. Write the exterior angle theorem. How can it be used to help find the angles?

5. What is true of all the possible triangles in Problem 3? Prove your answer.

6. Repeat Problem 1 with inscribed triangles such that the circle's center is inside the triangle. (See circle 3.)

7. Repeat Problem 1 with inscribed triangles such that the circle's center is outside the triangle. (See circle 4.)

Discussion and Wrap up

Questions and key points

Which triangles are impossible in Problem 1? Why?

In Problem 1, what is the smallest possible angle at the vertex that is at the circle's centre? Explain

Explain why the angle you found in Question B is the key to Problem 1.

How do you use the isosceles triangle theorem to find the other two angles of the triangles in Problem 1?

What is true of all the triangles in Problem 6? All the triangles in Problem 7?

Where to from here?

Now that you have established how triangles must form within circles, you are ready to explore other properties of circles and line segments known as chords. This will be the topic of the next few activities.

GEOMETRIC REASONING LESSON

8

The Intercepted Arc

A.C. Level:	10A
A.C. Ref No's:	Prove and apply angle and chord properties of circles ACMMG272
A.C. Substrands	Geometric Reasoning

Outcome

At the end of the activity students will know,

- The connection between angles formed at the centre of a circle and those formed on the circumference ending at the same points

Students will be able to

- Find the size of an angle either at the centre or on the circumference, given the value of the other angle

- Be able to prove why the relationship is always true

Materials Required

Calculator, Mathomat, Circle Geoboard

Definitions

A central angle is one with its vertex at the centre of the circle.
An inscribed angle is one with its vertex on the perimeter of the circle.
The arc intercepted by an angle is the part of the circle that is inside the angle.
The measure of an arc in degrees is the measure of the corresponding central angle.

1   Use the definitions above to fill in the blanks:

In the figure below

Angle APB is _____

Angle AOB is _____

Length AB is _____

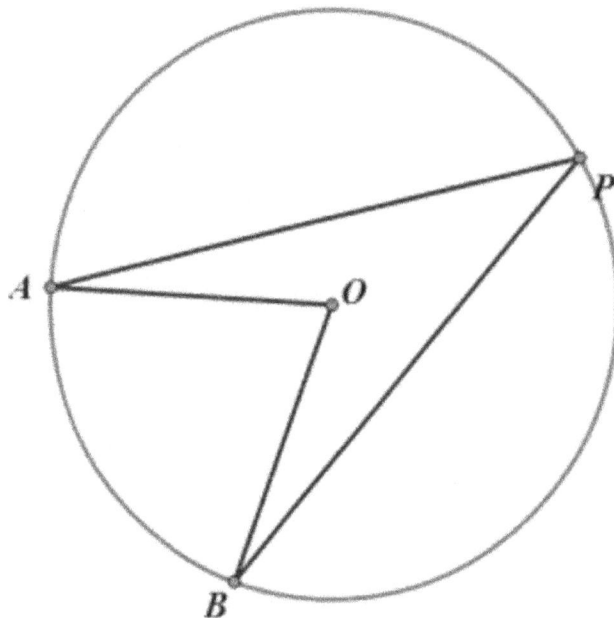

2   In the figure below which arc is intercepted by angle AOB?

Which arc is intercepted by angle APB?

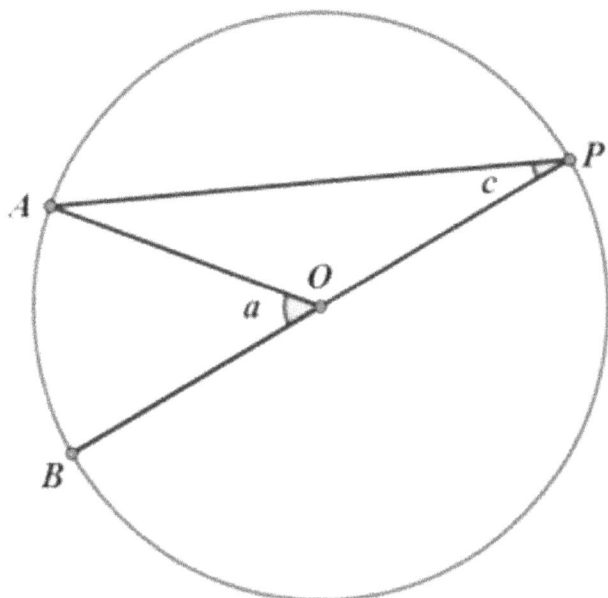

3.   If angle a = 50°, what is angle c? Explain.

4.   In general, what is the relationship between the angle at a and the angle at c? Explain.

5   In the next figure, if angle AOB = 140° and angle a = 50°, what is?
     What is angle c?
     What is angle d?
     What is angle APB?

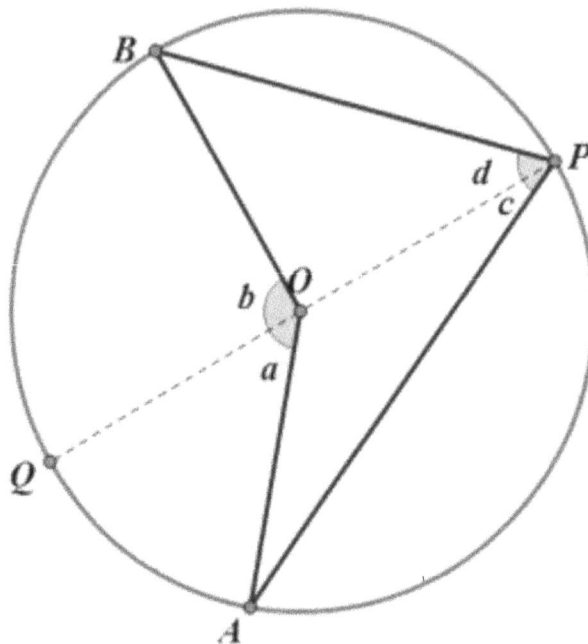

6.   Repeat Problem 5 with angle AOB = 140° and three different values for
     angle a.
     Arrange your results in a table.

7.   What is the relationship between angle APB and angle AOB? Explain.

8.   Write a sentence about the relationship between an inscribed angle and
     the corresponding central angle. If you do this correctly, you have stated
     the inscribed angle theorem.

9.   On a separate sheet, use algebra to prove the inscribed angle theorem for
     the case illustrated in Problem 5.

10. There is another case for the figure in Problem 5, where 0 is outside of
     angle APB. On a separate sheet, draw a figure and write the proof for
     that case.

A. Find the measure of angles a, b, c and d in the figure at right. Explain.

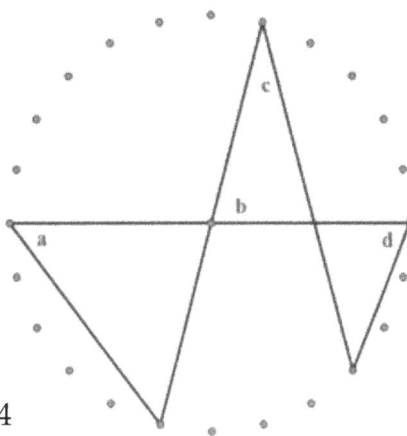

Figure 4

B. What are the interior angles in the triangles below?

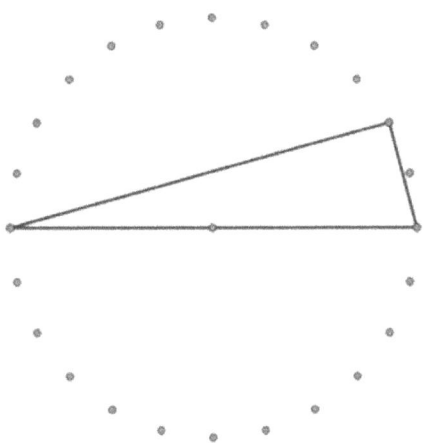

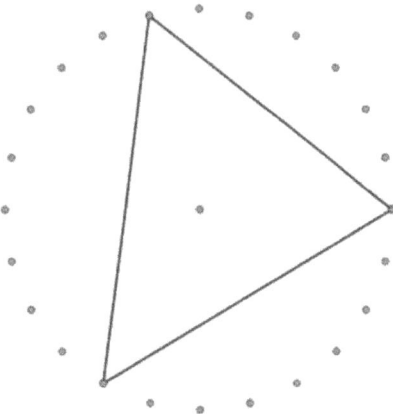

d. What are tile measures of tile central angle, the corresponding inscribed angle, and the intercepted arc in tile figure at right?

e. Inscribe a triangle in the circle geoboard so that its interior angles are 45°, 60°, and 75°.

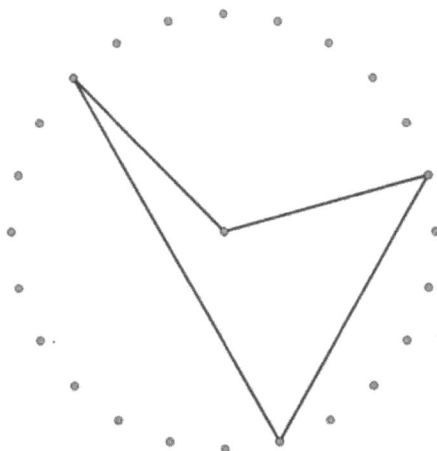

Discussion

Questions and key points

Complete the following statement:

The angle at the centre of a circle is _____ the size of the angle at the circumference when both angles start and finish at the same points.

When the angle at the centre of a circle is 180°,

a)   the two radii forming the angle combine to make the _____ of the circle.

b)   What is the size of the angle on the circumference?

c)   Does it matter where the vertex of this angle is? Explain your answer, including diagrams.

Where to from here?

The special case where the angle at the centre is 180° has some useful applications. It can be used to construct circles using a number of non standard tools. Investigate the Carpenters Construction of a Circle to see how this is done and then write up an explanation of why it can be a useful skill to have.

GEOMETRIC REASONING LESSON

9

Tangents and Inscribed Angles

A.C. Level:	10A
A.C. Ref No's:	Prove and apply angle and chord properties of circles ACMMG272
A.C. Substrands	Geometric Reasoning

Outcome

At the end of this activity:

Students will know

- The definition of a tangent to a circle

Students will be able to

- Connect what they know about intercepted angles and extend this knowledge to the case of a tangent to the circle

1. In the figure at right, find the intercepted arc and the measure for each of the following angles.

a.   Angle QPA intercepts arc

Angle QPA=

b.   Angle QPB intercepts arc

Angle QPB=

c.   Angle QPC intercepts arc

Angle QPC=

d.   Angle QPD intercepts arc

Angle QPD=

e.   Explain how you found the angle measures.

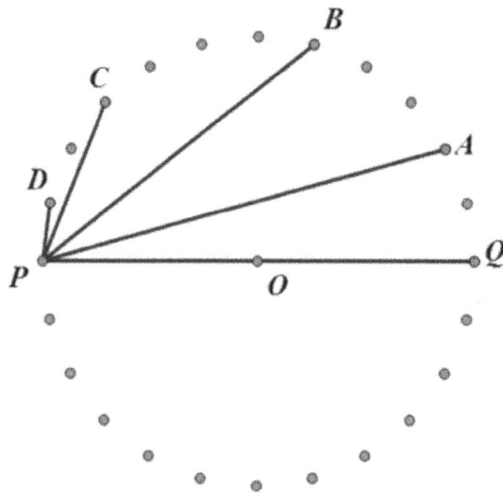

2. Segment PT is tangent to the circle (it touches it in exactly one point, P, which is called the point of tangency).

a. What arc is intercepted by LQPT?

b. What is the measure of Angle QPT?

3. Important: A segment that is tangent to a circle is _____ to the radius at the point of tangency.

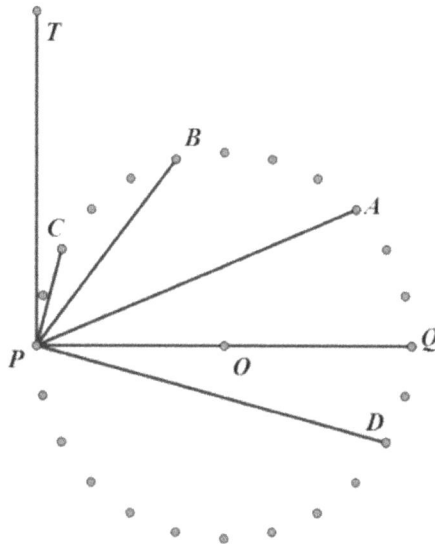

4. In the figure above, PT is tangent to the circle. Find the intercepted arc and the measure for each of the following angles.

Explain on a separate sheet how you found the angle measures.

a. TPA intercepts arc

Angle TPA=

c. TPC intercepts arc

Angle TPC=

b. TPB intercepts arc

Angle TPB=

d. TPD intercepts arc

Angle TPD=

5 Use Problem 4 to check that the inscribed angle theorem still works if one side of the angle is tangent to the circle.

Discussion

Questions and key points

The intercepted arc theorem, from Lesson 8, stated that the angle at the centre of a circle was double that of the angle on the circumference if both angles start and finish at the same points.

This lesson extends the concept outside of the circle along tangent lines.

Does the theorem still work? Explain why in your own words.

Where to from here?

Does the theorem still hold if the point T is drawn on the opposite side of the point of contact, that is underneath point P rather than above it? Explore the possibility using the diagrams

GEOMETRIC REASONING LESSON

10

Soccer Angles

A.C. Level:	10A
A.C. Ref No's:	Prove and apply angle and chord properties of circles ACMMG272
A.C. Substrands	Geometric Reasoning

Outcome

At the end of this activity:

Students will know:

- The relationship between an arc and the maximum angle that can be subtended by it.

Students will be able to:

- Use their knowledge to explore and solve how angle size changes with arc size.

Material Required

Cardboard, unlined paper, Soccer Angles Worksheet, Soccer Circles Worksheet, scissors, straight pins, Mathomat, Geometers Sketchpad.

Soccer goals are 8 yards wide. Depending on where a player is standing, the angle she makes with the goalposts could be larger or smaller. We will call this angle the shooting angle. In this figure, the space between G and H is the goal, P is the player, and L GPH is the shooting angle.

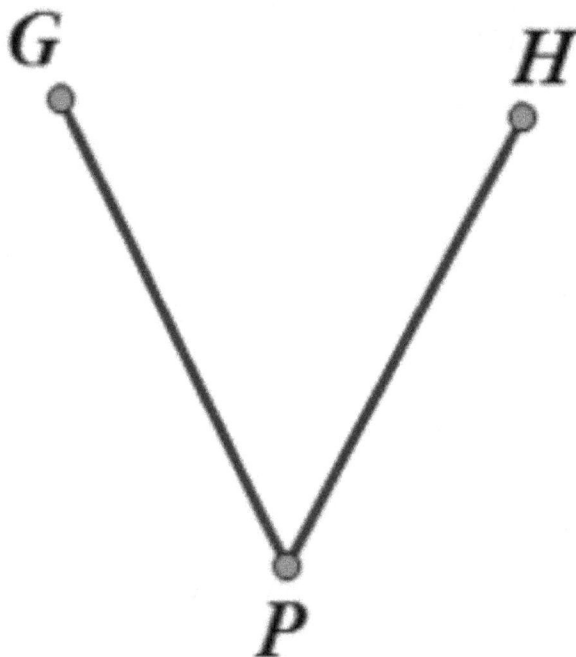

1.   Place a piece of paper on a cardboard backing. Place two pins as goalposts five centimetres apart, centred, near the top of the page. Label the goalposts G and H.

2.   Cut out these angles from the Soccer Angles Worksheet: 20°, 30°, 40°, 60°, 90°, 120°. (Alternatively use the Mathomat to construct these angle on paper.)

3.   Using the paper angle and the pins, find the location of all points that are the vertices of a shooting angle equal to 40°.

4.   Repeat Problem 3 with the other five angles cut from the Soccer Angles Worksheet, indicating clearly which points correspond to each shooting angle.

        If you did this correctly, you should have found that the locations form arcs of circles that pass through G and H. You may remove the pins and ask your teacher for the Soccer Circles Worksheet, where the circles are drawn very accurately for you.

5.   Label each arc with the measure of the corresponding shooting angle.

6.   Which of the arcs is a half-circle?

7.   The centre of each of the six arcs is marked. Label each centre so you will know which shooting angle it belongs to. (Use the notation C20, C40 , and so on.)

8.   Imagine a player is standing at C40, the centre of the 40° circle. What is the shooting angle there?

9.   Find the shooting angle for a person standing at the centre of each of the remaining five circles. What is the pattern?

10.   Find the centre of the 45° circle without first finding points on the circle.

Discussion

The vertices of the 40° shooting angles all lie on a circle. They are the vertices of inscribed angles. Where is the arc that is intercepted by those angles?

How is the inscribed angle theorem helpful in understanding what happens in this activity?

For Questions C-E, refer to the Soccer Discussion Sheet.

Imagine a player is running in a direction parallel to the goal, for example, on line Lo. Where would he get the best (greatest) shooting angle? (Assume the player is practicing, and there are no other players on the field.)

Imagine a player is running in a direction perpendicular to the goal, on a line that intersects the goal, for example, L_2
Where would she get the best shooting angle?

Imagine you are running in a direction perpendicular to the goal, on a line that does not intersect the goal, such as L_3 or L_4.
Where would you get the best shooting angle?

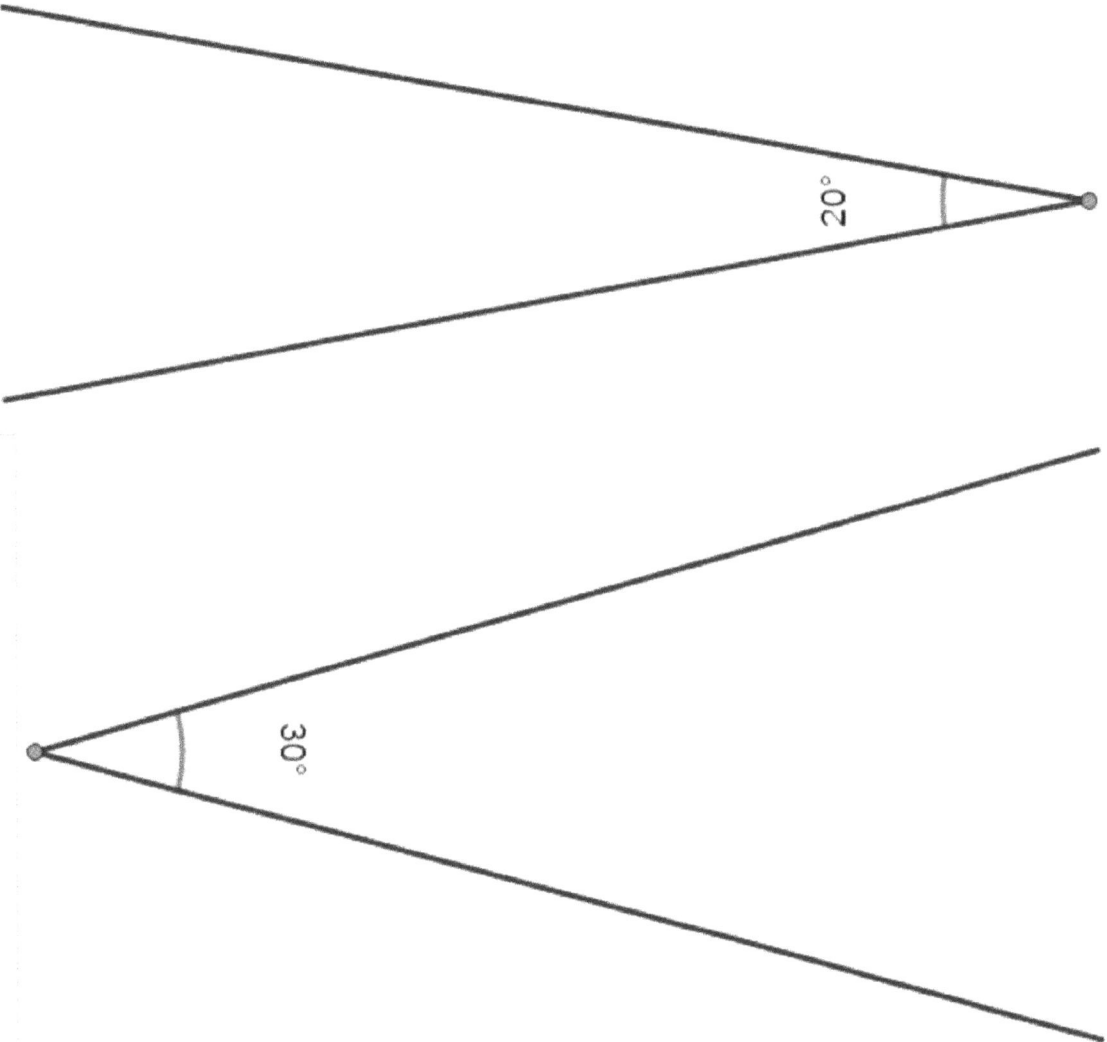

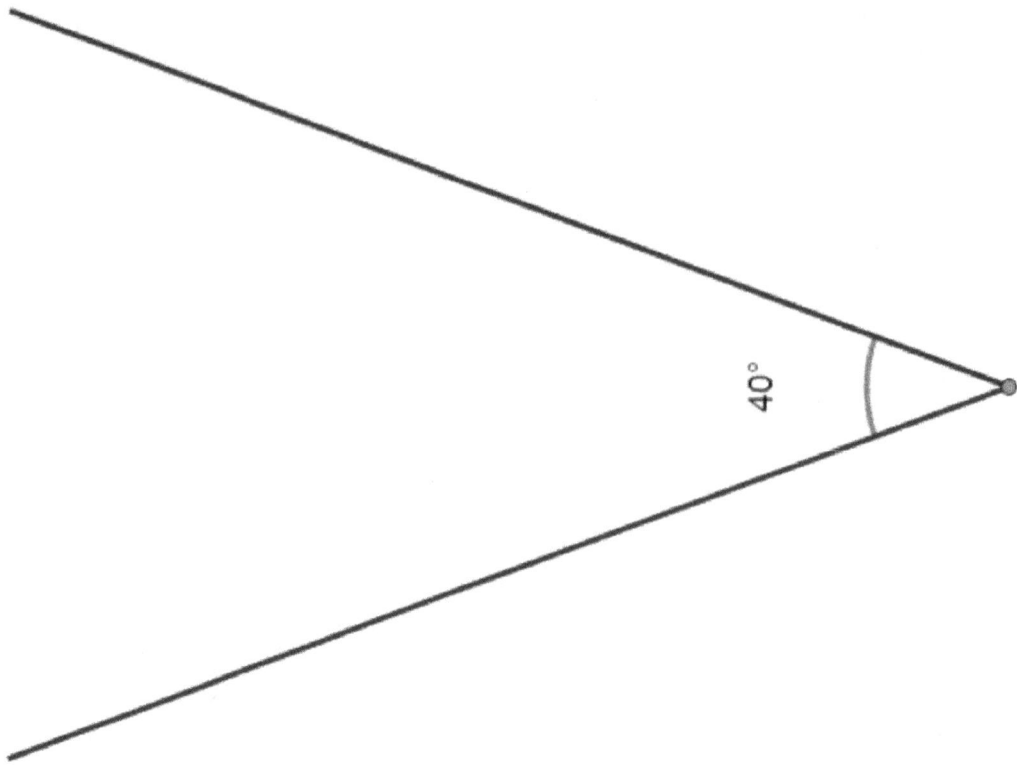

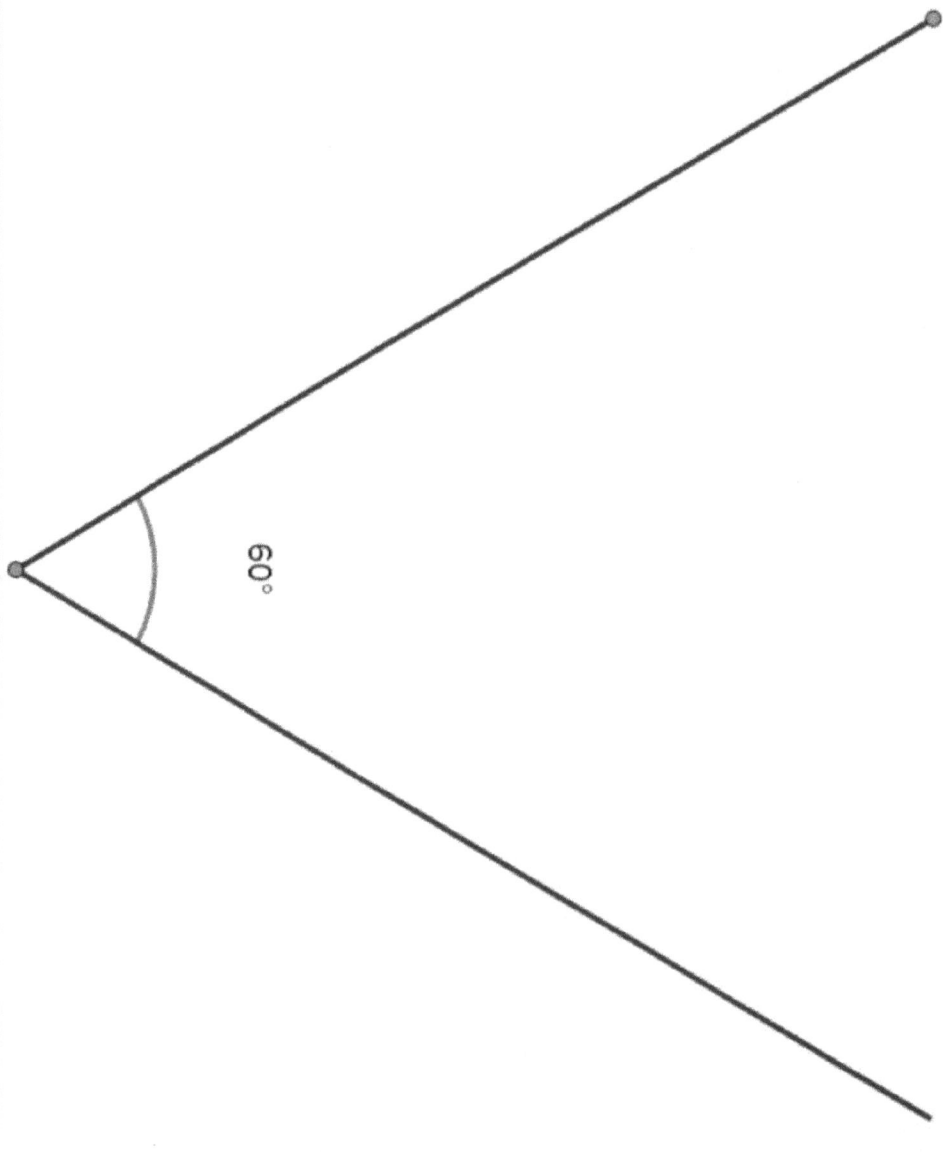

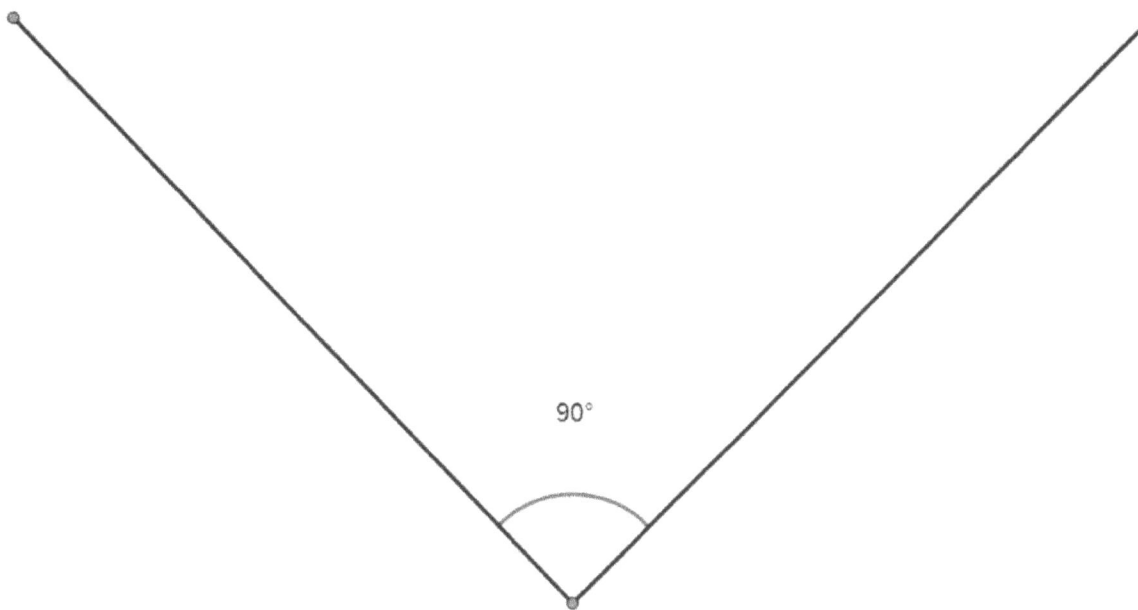

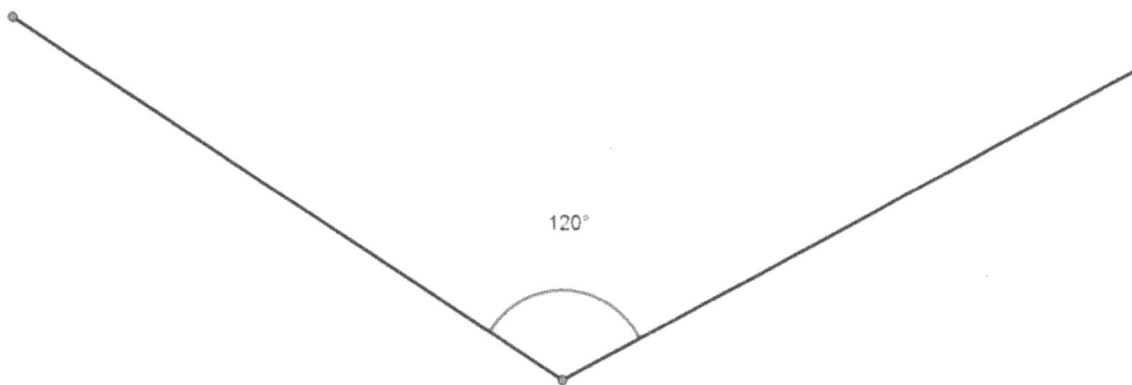

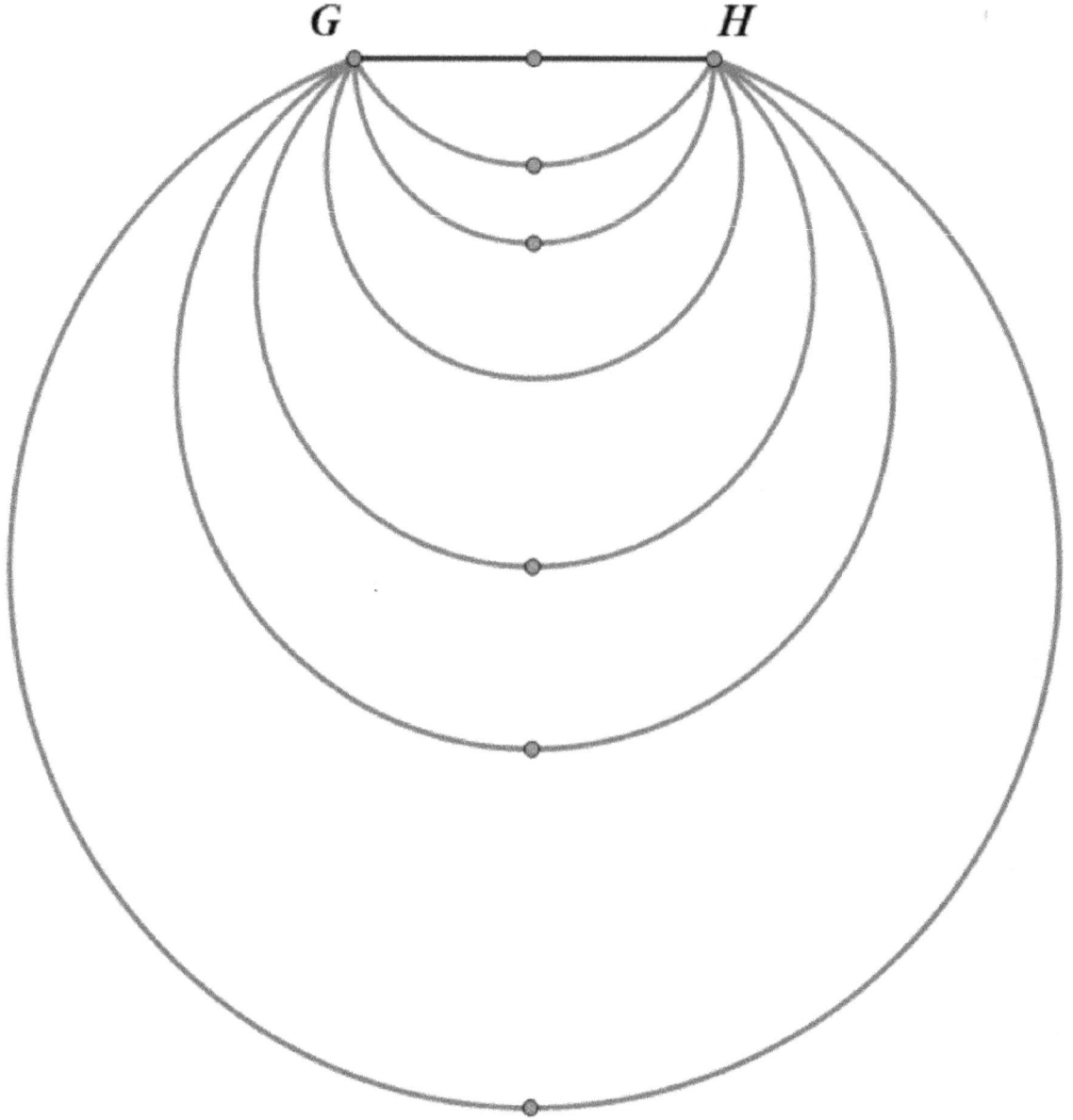

GEOMETRIC REASONING LESSON

11

Penrose Tiling

As you have seen, pentagons do not tessellate either regularly or semi-regularly.

So, what sort of pattern can be formed by starting with a pentagon and repeating it?

Begin, as in the diagram, with a single pentagon. (Length of a side is 17 mm Shape 10).

A.C. Level:	8 to 10
A.C. Ref No's:	This concept is not specified in the Australian Curriculum. It is connected to the properties of triangles and is included here as an extension activity.
A.C. Substrands	Geometric Reasoning

Outcome

At the end of this activity:

Students will know:

•   How to calculate the perimeters and areas of various closed polygons.

Students will be able to:

•   Describe and compare polygons of different shapes when given values for perimeter and area

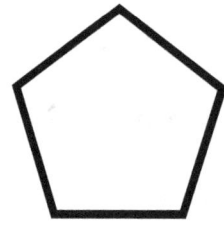

Use each of the sides of the pentagon as mirrors.
The first pentagon is now at the centre of five others.
There are V shaped gaps between each of these new pentagons.
The next step is to reflect this new composite shape another five times.

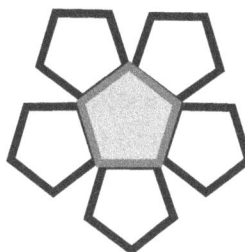

Continuing this process produces what looks like a semi-regular tessellation after three iterations.
After 4 iterations though, the pattern becomes more complicated. This is where using ICT becomes the more powerful approach.

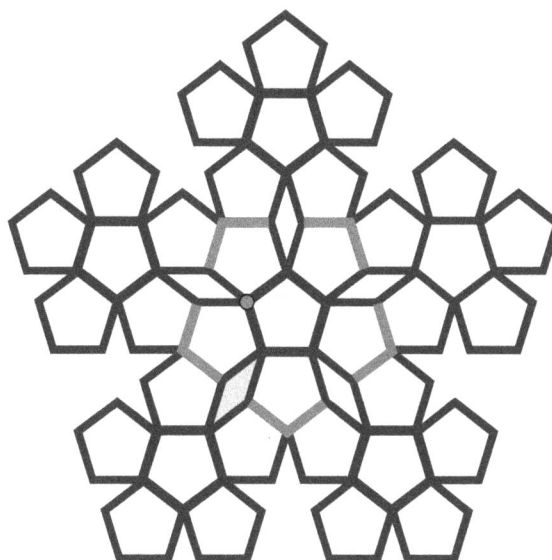

Here we use Sketchpad / Web Pages or Virtual Mathomat.

The pattern has now become much more detailed.
To make it clearer where the previous iteration of the tiling is situated, press the Match button to slide it over. You can return the smaller pattern to its original position by clicking on the Reset button.

It should be clear now that the pattern is not semi-regular but is producing further shapes, Stars and partial stars are shaded.

The pattern does however fill the entire space shich is the definition of a tiling pattern.

This particular pattern is known as Penrose Tiling. It was first discovered by Professor Roger Penrose, a British physicist.

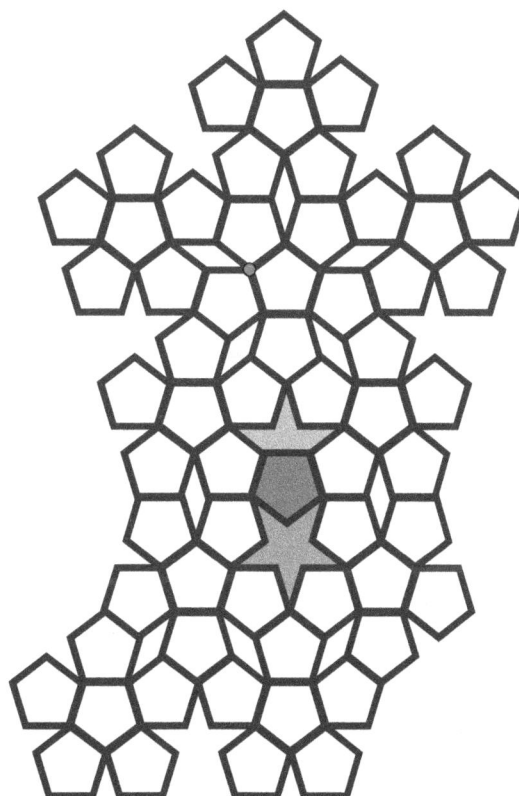

Tessellation

A tessellation is a tiling pattern that completely covers a plane.

Regular tessellations use only one regular polygon to create the pattern.

Semi-regular tessellations use more than one shape in the pattern but the basic arrangement of polygons around a point is always the same.

Non-periodic tilings have shapes that are not regular and also combine in different ways around different points in the pattern.

3 Types of Penrose Tiling called P1, P2 and P3

Roger Penrose investigated non-periodic tiling during the 1970's. There are three different Penrose

Tiling patterns, named P1, P2 and P3.

The components of each of these tilings are:

P1: Five-fold (Pentagonal Symmetry)

 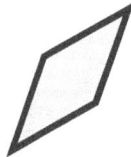  

Pentagon   Rhombus   Boat   Star Pentagram

P2: Kite (72°, 72°, 72° and 144°) and Dart (72°, 36°, 36°, 216°)

 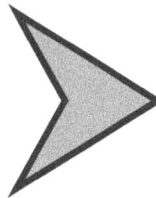

P3: Fat (72° and 108°) and Thin (36° and 144°) Rhombi

 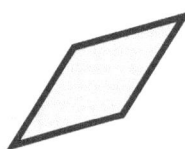

This is the version used in the façade at Storey Hall.

Discussion

Questions and key points

Are you convinced that the tiling pattern described in this activity:

1)   Will cover the plane with only the four shapes described

2)   It will never repeat itself

Discuss this with your classmates to make sure all of you are convinced.

What about for the other two types of Penrose Tiling: P2 and P3?
Explore the behaviour of these shapes in a similar way to how we have
worked with the P1 set.

Where to from here?

Tessellation, the covering of a plane surface using a limited number of shapes,
has been used in art for thousands of years. With the advent of computers
these techniques are finding ever more interesting uses. Modern animation
for tv, movies and games are some examples. Investigate how tiling similar
techniques have developed in recent years. You might want to start by
watching the short video from one of the Pixar Researchers.

http://ed.ted.com/lessons/pixar-the-math-behind-the-movies-tony-derose

Appendix: Why there are 360 degrees in a circle.

The ancient Sumerians and Babylonians had a numbering system based on the number 60.

They also knew that the perimeter of a hexagon was six times the radius of the circle into which the hexagon could just be fitted.

They then split the hexagon into six equilateral triangles and used this as their basic component of arc.

Finally, they decided to make the angle formed by an equilateral triangle equal to 60° which matches up with their numbering system. Having made this decision the number of degrees in a complete circle is 6 times 60° or 360°.

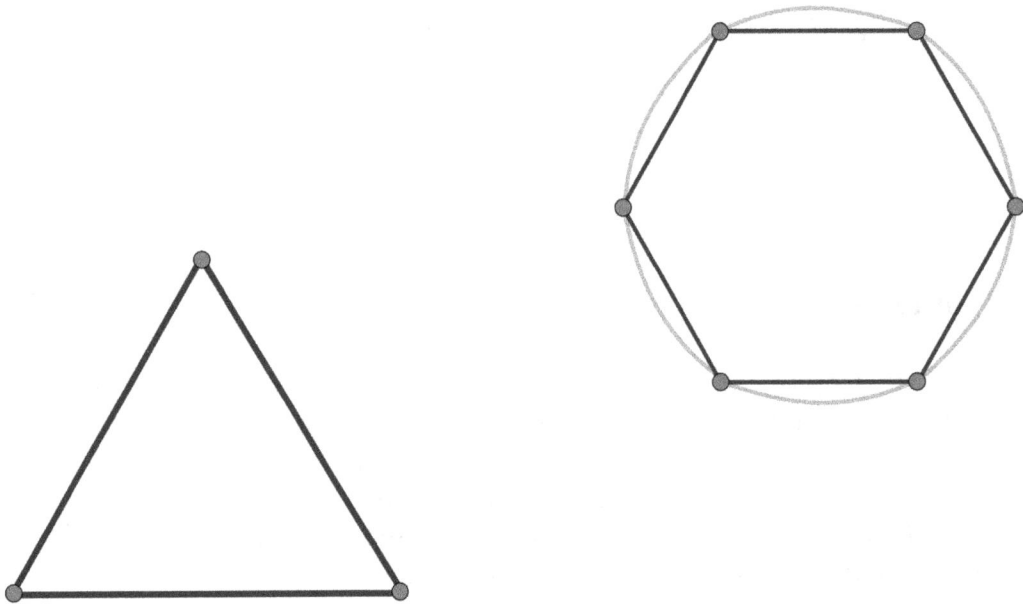

Constructing Degree Markings on a Compass

Take a piece of rope or string 540 mm long. Mark off 1.5 mm lengths along the entire length of rope. Join the ends of the rope together and form it into a circle. Draw the circumference of the circle onto something hard and also mark the centre of the circle. Use a straight edge to draw radii connecting the centre of the circle to each of the unit marks on the circumference.

Now by drawing smaller circles inside the original it is possible to produce a template for compasses of smaller sizes.

Measurement				
Lesson Name and Number	**A.C. Level**	**A.C. Substrands**	**A.C. Ref No's**	**Content Descriptor**
1 Measuring Rectangles	5	Measurement	ACMMG109	Calculate the perimeter and area of rectangles using familiar metric units
2 Measuring Many Sided Figures	6	Measurement	ACMMG137	Solve problems involving the comparison of lengths and areas using appropriate units
3 Area Connections: Rectangles, Triangles and Parallelograms	7	Measurement	ACMMG 159	Establish the formulas for areas of rectangles, triangles and parallelograms and use these in problem solving.
4 Diagonals of Squares: A Dangerous Idea	8	Real Numbers	ACMNA186	Investigate the concept of irrational numbers, including π
5 Three Quadrilaterals: Perimeter and Area	8	Real Numbers	ACMMG197	Find perimeters and areas of parallelograms, rhombuses and kites
6 π	8	Real Numbers	ACMNA186	Investigate the concept of irrational numbers, including π

Shape				
Lesson Name and Number	**A.C. Level**	**A.C. Substrands**	**A.C. Ref No's**	**Content Descriptor**
1 Drawing Solids	5	Shape	ACMMG111	Connect three dimensional objects with their nets and other two dimensional representations
2 Prisms and Pyramids	6	Shape	ACMMG140	Construct simple prisms and pyramids
3 Constructor Challenge	7	Shape	ACMMG161	Draw different views of prisms and solids formed from combinations of prisms

Location and Transformation

Lesson Name and Number	A.C. Level	A.C. Substrands	A.C. Ref No's	Content Descriptor
1 Map Grids	5	Location and Transformation	ACMMG113	Use a grid reference system to describe locations. Describe routes using landmarks and directional language
2 Wallpaper Patterns	5	Location and Transformation	ACMMG114	Describe translations, reflections and rotations of two-dimensional shapes. Identify line and rotational symmetries

Pythagoras and Trigonometry

Lesson Name and Number	A.C. Level	A.C. Substrands	A.C. Ref No's	Content Descriptor
1 Right Angled Triangles are Special	9	Pythagoras and Trigonometry	ACMMG222	Investigate Pythagoras' Theorem and its application to solving simple problems involving right angled triangles
2 Taxi Cab Distances	9	Linear and non-linear relationships	ACMNA214	Find the distance between two points located on a Cartesian plane using graphical and algebraic techniques, including graphing software.
3 Right Angled Ratios	9	Pythagoras and Trigonometry	ACMMG223	Use similarity to investigate the constancy of the sine, cosine and tangent ratios for a given angle in right-angled triangles
4 Exploring Patterns in Triangle Ratios	9	Pythagoras and Trigonometry	ACMMG224	Apply trigonometry to solve right-angled triangle problems
5 A New Area Formula	10 A	Pythagoras and Trigonometry	ACMMG273	Establish a formula for area using the sine ratio.
6 The Sine Rule	10 A	Pythagoras and Trigonometry	ACMMG273	Establish the sine rule for any triangle and solve related problems
7 The Cosine Rule	10 A	Pythagoras and Trigonometry	ACMMG273	Establish the cosine rule for any triangle and solve related problems
8 The Unit Circle	10 A	Pythagoras and Trigonometry	ACMMG274	Use the unit circle to define trigonometric functions, and graph them with and without the use of digital technologies

Geometric Reasoning				
Lesson Name and Number	**A.C. Level**	**A.C. Substrands**	**A.C. Ref No's**	**Content Descriptor**
1 Observing Angles around a Point	5	Geometric Reasoning	ACMMG112	Estimates, measures and compares angles using degrees. Constructs angles using both a 180° and 360° protractor.
2 Angle Measurement	5	Geometric Reasoning	ACMMG112	Estimates, measures and compares angles using degrees. Constructs angles using both a 180° and 360° protractor.
3 Clock Angles	6	Geometric Reasoning	ACMMG141	Investigate, with and without digital technologies, angles on a straight line, angles at a point and vertically opposite angles. Use results to find unknown angles. Define acute, obtuse, reflex and right angles. Identify that angles have arms and a vertex and that the size is the amount of turn required for one arm to coincide with the other. The angle is measured in degrees with a protractor
4 Angles of Mathomat Polygons	7	Geometric Reasoning	ACMMG163	Demonstrate that the angle sum of a triangle is 180° and use this to find the angle sum of a quadrilateral
5 Classifying Angles	7	Geometric Reasoning	ACMMG165	Classify triangles according to their side and angle properties of scalene, isosceles, right-angled and obtuse-angled triangles
6 The Exterior Angle Theorem	7	Geometric Reasoning		This concept is not specified in the Australian Curriculum. It is connected to the properties of triangles and is included here as an extension activity.
7 Angles and Triangles in a Circle	10A	Geometric Reasoning	ACMMG272	Prove and apply angle and chord properties of circles
8 The Intercepted Arc	10A	Geometric Reasoning	ACMMG272	Prove and apply angle and chord properties of circles
9 Tangents and Inscribed Angles	10A	Geometric Reasoning	ACMMG272	Prove and apply angle and chord properties of circles
10 Soccer Angles	10A	Geometric Reasoning	ACMMG272	Prove and apply angle and chord properties of circles
11 Penrose Tiling	8 to 10	Geometric Reasoning		This concept is not specified in the Australian Curriculum. It is connected to the properties of triangles and is included here as an extension activity.